Synthesis Lectures on Information Concepts, Retrieval, and Services

Editor
Gary Marchionini, *University of North Carolina at Chapel Hill*

Synthesis Lectures on Information Concepts, Retrieval, and Services is edited by Gary Marchionini of the University of North Carolina. The series will publish 50- to 100-page publications on topics pertaining to information science and applications of technology to information discovery, production, distribution, and management. The scope will largely follow the purview of premier information and computer science conferences, such as ASIST, ACM SIGIR, ACM/IEEE JCDL, and ACM CIKM. Potential topics include, but are not limited to: data models, indexing theory and algorithms, classification, information architecture, information economics, privacy and identity, scholarly communication, bibliometrics and webometrics, personal information management, human information behavior, digital libraries, archives and preservation, cultural informatics, information retrieval evaluation, data fusion, relevance feedback, recommendation systems, question answering, natural language processing for retrieval, text summarization, multimedia retrieval, multilingual retrieval, and exploratory search.

Information Architecture: The Design and Integration of Information Spaces
Wei Ding and Xia Lin
2009

Reading and Writing the Electronic Book
Catherine C. Marshall
2009

Hypermedia Genes: An Evolutionary Perspective on Concepts, Models, and Architectures
Nuno M. Guimarães and Luís M. Carrico
2009

Understanding User-Web Interactions via Web Analytics
Bernard J. (Jim) Jansen
2009

XML Retrieval
Mounia Lalmas
2009

Faceted Search
Daniel Tunkelang
2009

Introduction to Webometrics: Quantitative Web Research for the Social Sciences
Michael Thelwall
2009

Exploratory Search: Beyond the Query-Response Paradigm
Ryen W. White and Resa A. Roth
2009

New Concepts in Digital Reference
R. David Lankes
2009

Automated Metadata in Multimedia Information Systems: Creation, Refinement, Use in Surrogates, and Evaluation
Michael G. Christel
2009

Visual Information Retrieval using Java and LIRE

Mathias Lux and Oge Marques

ISBN: 978-3-031-01154-2   paperback
ISBN: 978-3-031-02282-1   ebook

DOI 10.1007/978-3-031-02282-1

A Publication in the Springer series
SYNTHESIS LECTURES ON INFORMATION CONCEPTS, RETRIEVAL, AND SERVICES

Lecture #25
Series Editor: Gary Marchionini, *University of North Carolina at Chapel Hill*
Series ISSN
Synthesis Lectures on Information Concepts, Retrieval, and Services
Print 1947-945X   Electronic 1947-9468

Visual Information Retrieval using Java and LIRE

Mathias Lux
Alpen Adria Universität Klagenfurt, Klagenfurt, Austria

Oge Marques
Florida Atlantic University, Boca Raton, Florida

SYNTHESIS LECTURES ON INFORMATION CONCEPTS, RETRIEVAL, AND SERVICES #25

ABSTRACT

Visual information retrieval (VIR) is an active and vibrant research area, which attempts at providing means for organizing, indexing, annotating, and retrieving visual information (images and videos) from large, unstructured repositories.

The goal of VIR is to retrieve matches ranked by their relevance to a given query, which is often expressed as an example image and/or a series of keywords. During its early years (1995-2000), the research efforts were dominated by content-based approaches contributed primarily by the image and video processing community. During the past decade, it was widely recognized that the challenges imposed by the lack of coincidence between an image's visual contents and its semantic interpretation, also known as *semantic gap*, required a clever use of textual metadata (in addition to information extracted from the image's pixel contents) to make image and video retrieval solutions efficient and effective. The need to bridge (or at least narrow) the semantic gap has been one of the driving forces behind current VIR research. Additionally, other related research problems and market opportunities have started to emerge, offering a broad range of exciting problems for computer scientists and engineers to work on.

In this introductory book, we focus on a subset of VIR problems where the media consists of images, and the indexing and retrieval methods are based on the pixel contents of those images—an approach known as *content-based image retrieval* (CBIR). We present an implementation-oriented overview of CBIR concepts, techniques, algorithms, and figures of merit. Most chapters are supported by examples written in Java, using Lucene (an open-source Java-based indexing and search implementation) and LIRE (Lucene Image REtrieval), an open-source Java-based library for CBIR.

KEYWORDS

information retrieval, image search, image retrieval, visual search, indexing, visual descriptors, image processing, Java

To my kids, Annika, Emma, Leni, and Samuel, and to my wife, Daniela.

– Mathias Lux

To my son Nicholas.

– Oge Marques

Contents

Preface

Since 2006, LIRE has been an ongoing thread through my academic career and my public profile. Funny enough, this has never been my intention. The first steps were taken by integrating content-based retrieval functions into Caliph&Emir, a toolset I'd developed for creating and retrieving MPEG-7 metadata, just to back my idea of metadata-based retrieval, which I pursued during my Ph.D. years. At that time I took the MPEG XM software and adapted three of the visual descriptors to Java. During a demonstration of the functionality of Caliph&Emir in the course of presenting my Ph.D. work at Graz University of Technology, a friend and colleague, Roman Kern—with whom I worked on a Lucene search project for a large company—approached me. He asked for code to integrate image retrieval functionality into his web portal. That was the time I created a first version of a library that uses Lucene for image retrieval, based on the three descriptors from Caliph&Emir. In search for a name we joined the description of the library—Lucene Image REtrieval—with our love for holidays in Italy, Italian food, and its former currency…and came up with LIRE.

I released LIRE under GPL in 2006 on sourceforge.net. By that time I was already known within a certain research community for developing and maintaining Caliph&Emir, so I just attached LIRE there as a sub project and things went on for a while. Unnoticed by me, LIRE went from sub project to main driving force. People started to contact me for good tips on deployment, potential bugs, code contributions, suggestions, etc. Also, the research community—in need for a tool for teaching content-based image retrieval and a baseline for testing common features against new ones—adopted LIRE.

A boost in importance of LIRE was the contribution of CEDD and FCTH by my colleague and friend Savvas A. Chatzichristofis. His contributions made LIRE a library featuring not only common, old descriptors, but also newly created, state-of-the-art ones. Our joint ACM Multimedia 2008 paper on LIRE has been cited more than 100 times since its publication. During the same ACM Multimedia 2008 conference in Vancouver I was approached by Marco Bertini, whose work I knew well from his research papers, who told me how much he liked LIRE and appreciated my work. Then I finally realized that LIRE had outgrown Caliph&Emir and I re-focused my attention on LIRE.

Oge Marques already knew me from my work on Caliph&Emir and LIRE when he met the head of my department—Laszlo Böszörmenyi—at a conference in Innsbruck in March 2008 and they arranged for a visit from Oge to Klagenfurt in the summer of that year. Since we first met in May 2008, Oge and I have kept an ongoing professional relationship which has produced joint research efforts, several co-authored papers, co-chaired events, and student exchanges. This successful collaboration bore the seed for this book.

Now LIRE is a stable project, with a good amount of academic backing. LIRE has been and is deployed currently in several small- and large-scale industry projects and is a direct competitor to commercial products. LIRE has been successfully employed in teaching and research, and has been considered by many experts as a great tool to get started with content-based image retrieval. It became a constant factor in my life, with a significant percentage of time perennially spent on bug fixing, developing, helping other developers, or maintaining documentation and code several times a week. Finally, thanks to this book, it is now the only content-based image retrieval open source library, that I know of, for which a book has been written.

Mathias Lux
Klagenfurt, December 2012

When Mathias and I first started discussing the possibilities of writing a book together. I had a number of goals in mind. I wanted the final result to be a concise technical guide to the field of visual information retrieval with a clear emphasis on practical, hands-on aspects that would allow our readers to build their own VIR solutions using the Java programming language and the LIRE library for content-based image retrieval. I expected that the work would summarize some of the latest and most relevant developments in the field (even if that meant rendering my previous book on the topic all but obsolete). I wished that the book would do justice to the richness of features and functionalities available through LIRE and explained its architecture in enough detail. Moreover, I hoped that in our attempt to focus primarily on design and implementation aspects of VIR solutions we would not sacrifice the explanation of basic concepts, the appropriate mathematical rigor expected when presenting key topics in the discipline, and an overall broad understanding of the field and its challenges and opportunities.

Now what was once a plan became reality. Here is the book. Did we succeed in making the book achieve its goals? You, dear reader, will be the judge.

A book is never complete, especially in a fast-changing field such as visual information retrieval. To keep up with the most relevant developments and updates in the field, the book's companion website (`http://www.lire-project.net/`) will stay updated with the latest version of LIRE and a list of references, datasets and tools, which may be of use to those conducting research in this area.

Oge Marques
Boca Raton, Florida, December 2012

Acknowledgments

We'd like to thank all the LIRE contributors, who have made this effort a true open source and community product: Anna-Maria Pasterk, Arthur Li, Arthur Pitman, Bastian Hösch, Benjamin Sznajder, Christian Penz, Christine Keim, Christoph Kofler, Dalibor Mitrovic, Dan Hanley, Daniel Pötzinger, David Zellhöfer, Fabrizio Falchi, Giuseppe Amato, Janine Lachner, Katharina Tomanec, Lukas Esterle, Manuel Oraze, Marian Kogler, Mario Taschwer, Marko Keuschnig, Rodrigo Carvalho Rezende, Roman Divotkey, and Roman Kern.

We'd especially like to thank our colleague and friend Savvas A. Chatzichristofis for his contributions to LIRE and the fruitful discussions and ongoing collaboration on content-based image retrieval and related topics.

This book could not have been produced without the vision of Joel Claypool, the support of our series editor, Gary Marchionini, the patient and friendly guidance and encouragement from our Executive Editor Diane Cerra, and the expert assistance with graphics and LaTeX issues from Clovis Tondo and his team.

Special thanks to Martina Steinbacher, whose unparalleled efficiency and professionalism can only be matched by her kindness, friendship, and encouragement.

Last, but not least, our biggest thanks to Laszlo Böszörmenyi, the head of the ITEC at Klagenfurt University, who was instrumental in establishing this partnership between the two authors. Thanks to his natural abilities of fostering relationships between people, supporting creative projects, and always keeping an open door, he greatly contributed to the successful completion of this project.

Mathias Lux and Oge Marques
December 2012

CHAPTER 1

Introduction

There has been an explosive growth in the use of visual digital media in everyday life. This trend has been fueled by many factors, among them advances in availability and prices for digital cameras, software for image manipulation processing and sharing, camera phones, and multimedia communication in social networks. Facebook alone is boasting more than 200 million image uploads every day.[1]

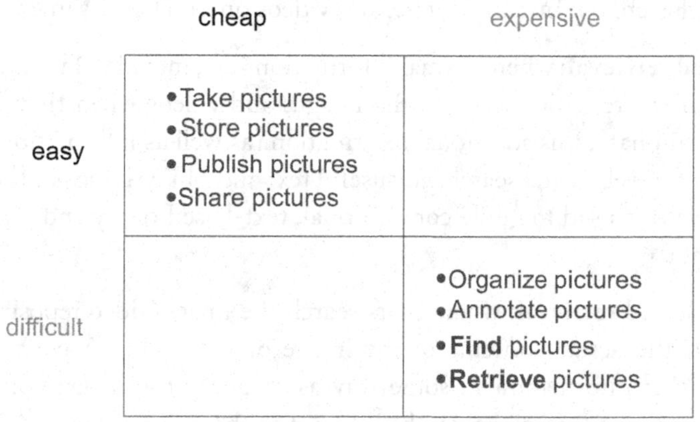

Figure 1.1: The "big mismatch": the four activities related to *image production* (top-left) are easy and inexpensive, whereas the four activities related to *image consumption* (bottom-right) are not.

Taking pictures, as well as storing, sharing, and publishing them, has never been so easy and inexpensive. On the other hand, finding the images we want and retrieving them remains, for the most part, difficult and time consuming. This "big mismatch" between two groups of tasks (Figure 1.1) summarizes some of the key motivations for the research work in the field of visual information retrieval (VIR) and the existence of this book. For the past 20 years, VIR researchers and developers have been working on solutions and algorithms that will improve the way we organize and annotate existing pictures and—when no such pictures are available—quickly find and retrieve the most relevant images for a certain query.

We can all relate to situations where we want a picture (e.g., a photo or clipart) to illustrate a presentation, report or handout and end up empty-handed in spite of our best efforts using general-

[1]http://www.quora.com/How-many-photos-are-uploaded-to-Facebook-each-day

purpose search engines (such as Google or Bing) and text-based queries. It was, in part, due to the shortcomings of text-based retrieval that a new approach for visual queries, based on visual properties of the desired image or video clip, emerged. It is called *content-based image retrieval* (CBIR) and it has recently been made visible to a large user audience thanks to Google Image Similarity Search,[2] introduced in 2009. Even though CBIR is unlikely to completely replace text-based queries, it might make for a convenient companion (in the case of web-based searches), as illustrated in Figure 1.2, which shows how a search for a yellow Ferrari can be achieved much more successfully by using the keyword "Ferrari" and a sample image containing a yellow Ferrari instead of the text query alone.[3]

Some of the most common ways of finding and retrieving images and videos include the following.

1. **Browsing**: where users browse through a collection of images and videos, and stop when they find the desired information. This is a very common way that people use to look at pictures on Flickr, Facebook, Instagram, etc., and videos on YouTube, Vimeo, and equivalent portals.

2. **Text-based retrieval**: where textual information (i.e., metadata) is added—either manually or using (semi-)automatic tools—to the images and videos when they are uploaded/stored. In the retrieval phase, this additional information (as well as information surrounding the visual asset, e.g., for web-based search, any useful text and HTML tags in the page where the image is embedded) is used to guide conventional, text-based query and search engines to find the desired data.

3. **Content-based retrieval**: where users search the image/video repository providing information about the actual contents of the image or video clip. A content-based search engine translates this information in some way as to query the image collection and retrieve the candidates that are more likely to be relevant to the user's request.

4. **Recommendation-based systems**: where, after watching or rating an image or video, additional related (and potentially relevant) images and videos are suggested to the user. These recommendations may be: (i) based on collaborative filtering algorithms (used by Amazon, YouTube, and Netflix, among others), which factor in the relationships between a certain image/video/movie and the past browsing/viewing history of users who have rated it positively; or (ii) driven by social aspects, such as the interestingness of a photo in Flickr, its popularity in Instagram, or Facebook friends' recommendations.

In this book we will focus primarily on content-based visual information retrieval and refer to its text-based counterpart whenever appropriate. Browsing will not be discussed any further, since it is covered in a different area of research, namely human-computer interfaces. For more information on image- and video-oriented recommendation-based systems, please see [47, 56, 61, 82, 83].

---

[2] http://www.google.com/insidesearch/features/images/searchbyimage.html

[3] The attentive reader will notice that this example has two shortcomings: (1) it neglects the fact that typing "yellow Ferrari" would probably solve the problem much more quickly, and without resorting to visual similarity search; and (2) it presumes that a suitable sample image is available. The latter is a frequent objection to the basic query-by-example (QBE) paradigm used in CBIR.

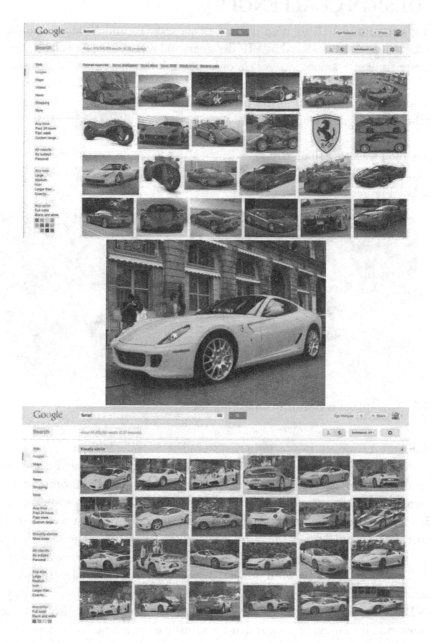

Figure 1.2: (Top): generic Google image search results for keyword "Ferrari"; (bottom): improved results (query image of a yellow Ferrari (center) **plus** keyword "Ferrari").

1.1 DESIGN CHALLENGES

Designing a successful VIR solution is not a trivial task. To underline the complexity we explain some of the main challenges faced by VIR designers and researchers, namely: (i) the need to capture and measure similarity among images (in order to rank the most relevant images to a query close to the top of the results list); (ii) the inherent difficulty in making VIR solutions work effectively in broad domains; (iii) the need to take users' intentions into account when designing VIR systems; and (iv) the *semantic gap* (along with other gaps).

Capturing and measuring similarity

There is an inherent difficulty in establishing and measuring degrees of similarity among images, as illustrated by Figures 1.3 and 1.4.

Figure 1.3: Example of two visually similar images.

In Figure 1.3 we see two pictures that would be considered (by most people) *similar* to each other mostly due to their *visual* similarities, e.g., the predominance of white and blue colors and an almost identical color layout. Although they would probably also share common keywords (or tags) if they were annotated (e.g., *snow*, *sky*, *trees*, *winter*, *Carinthia*), there is a fundamental difference in semantic meaning between the two images: if a user is searching for a picture of a ski station, the right-most image should satisfy the query, whereas the left-most image should not. Conversely, in Figure 1.4 we see two pictures that would be considered *semantically similar*, since they evoke similar concepts (for example, the keyword *flower*), despite their *visual* dissimilarities (the flowers are of different color, the overall brightness and contrast of each image are significantly different, etc.). From a VIR design viewpoint, trying to automatically determine, for an isolated query, whether the user is interested in retrieving images that are visually or semantically similar (or both) to a query image provided as an example is now tacitly recognized as an unsurmountable problem. One way to circumvent it consists in offering the user both options—often using words such as *related* and *similar*, which are interpreted by the VIR system as *semantically related* or *visually similar*, respectively.[4]

[4]This is used in Alipr (http://alipr.com/), for example.

Figure 1.4: Example of two semantically similar images.

The problem is aggravated when the VIR system designer is faced with the task of assigning a quantitative measure of (dis)similarity between images. As we shall see in chapter 3, VIR systems usually adopt distance metrics and other mathematical measures of divergence from information retrieval to act as a proxy for visual (dis)similarity, which have the advantage of producing precise values indicating how similar or dissimilar two sets of values (in this case, the feature vectors encoding the visual properties of the images being compared) are.

Narrow vs. broad domains
One of the major challenges faced by VIR systems is the ability to produce relevant results in broad domains (for example, large collections of personal photographs), due to the high variance of content (bright/dark, indoor/outdoor, faces/landscapes/monuments, etc.), heterogeneous semantics (the photos may have been taken for completely different purposes; if annotated, their tags would be significantly diverse), and uncontrolled image acquisition and production methods (different cameras and settings, flash/no-flash, etc.), among many others. Conversely, in narrow domains (for example, lung x-rays acquired by the same medical staff, using the same equipment, following a consistent—and properly documented—protocol) the VIR designer can leverage the advantages of: low variance of visual content, controlled and consistent vocabulary (if metadata, e.g., tags or equipment settings, can be used), availability of expert knowledge, and availability of ground truth data, e.g., examples of a cancerous lung vs. a healthy one [74].

Users' needs and intentions
Researchers in visual information retrieval have recently started to investigate the *context* of users' queries with the goal of uncovering the users' needs and intentions in a more clear and systematic way. In VIR systems, users might want to find images for completely different reasons, such as: to illustrate a presentation, to determine if a copyrighted image has been posted elsewhere on the web without permission, to gain knowledge on a place never visited, or simply to entertain themselves. Understanding the intentions of users may lead to better VIR systems in the future. User intentions

for searching images are discussed in [51], where also a taxonomy of user intentions for image retrieval is presented. A taxonomy on intention classes for online video search is discussed in [31]. An application of the research on user intentions for image search is discussed in [39], where the result view of Flickr is adapted to the automatically detected search intention class. More recently, Lux et al. have presented a public test data set (1,309 photos along with annotations specifying the intentions of the photographers when they took the pictures) for researchers interested in exploring these new avenues [52].

The semantic gap (and other gaps)

The semantic gap can be defined as "the lack of coincidence between the information that one can extract from the visual data and the interpretation that the same data have for a user in a given situation" [74]. It has been the topic of several entire papers (see for example [23] and [33]). The popularity of the phrase *semantic gap* inspired Deserno et al. to produce a paper [20] in which the most critical challenges in building CBIR solutions are expressed in terms of gaps, classified in four categories, as follows:

- Content-related:

  – Semantic gap: expresses the fact that humans understand images at a high semantic level that is hard to achieve using contemporary techniques for computational analysis of image contents.

  – Use context gap: addresses the broad vs. narrow domain aspects explained earlier in this section.

- Feature-related:

  – Extraction gap: refers to the amount of automation involved in the feature extraction process, from fully manual to fully automated.

  – Structure gap: expresses the differences in granularity of image object structures recognized by a CBIR system.

  – Scale gap: addresses the issue of whether visual details are processed at a single or multiple levels of scale.

  – Space plus time dimension gap: indicates the fact that most CBIR systems do not exploit all possible spatial and time dimensions available in the input data when computing features.

  – Channel dimension gap: denotes the related issue of how many channel dimensions of input data are used when computing features.

- Performance-related:

  – Application gap: addresses the difference between having a conceptual solution and deploying it, successfully, online.

– Integration gap: expresses how well the CBIR system is integrated with other information systems, e.g., patient care information systems (in the context of medical image search).

– Indexing gap: refers to the dramatic performance differences between a naïve linear search of the image collection to a highly optimized search strategy.[5]

– Evaluation gap: alludes to the need of evaluating contemporary CBIR systems with quantitatively with ground truth data, preferably from publicly available datasets.

• Usability-related:

– Query gap: refers to the number of query options available to the user, beyond query by text, e.g., query by example image, numerical specification of features, sketch, and so on.

– Feedback gap: expresses the level to which the CBIR system helps a user understand query results.

– Refinement gap: refers to the level to which the system helps the user to refine and improve query results.

By now you should understand the basic concepts and challenges behind the design of VIR systems. We encourage you to spend some time playing with contemporary web-based VIR solutions/prototypes, e.g., Google Image Search,[6] Alipr,[7] GIFT,[8] Incogna,[9] TinEye,[10] or pixolution[11] and use the opportunity to correlate your experience as a user with the concepts and design challenges mentioned in this section.

In the next section, we will take you on a tour of `LireDemo` and show you how to create your own desktop Java-based VIR solution.

1.2 GETTING STARTED WITH LIRE

In this section, we present the steps necessary to download, install, compile and run a sample desktop application, called `LireDemo`, based on the LIRE library. LireDemo is a friendly GUI-based utility created to allow Java and LIRE developers to quickly inspect the results of different visual features, indexing methods, and re-ranking algorithms applied to their document collections.[12] The `LireDemo` application will allow you to index the contents of an image collection based on their visual properties and use the resulting index to perform content-based visual queries using example images.

---

[5] See Chapter 4 for more details on indexing strategies and algorithms.
[6] http://images.google.com/
[7] http://alipr.com/
[8] http://gift.faikvm.com/
[9] http://www.incogna.com/
[10] http://www.tineye.com/
[11] hhttp://www.pixolution.de/
[12] Readers with Lucene experience will realize that LireDemo is to LIRE what Luke (Lucene Index Toolbox) [2] is to Lucene.

LIRE (Lucene Image REtrieval) is a content-based image retrieval library written entirely in Java [50]. LIRE extracts visual features from bitmap images and stores them in a Lucene [1] index for later retrieval. LIRE is targeted at developers and researchers who want to integrate content-based image retrieval (CBIR) features in their applications. Thanks to its simplicity and small footprint, LIRE provides an easy way to test the capabilities of CBIR approaches for different application domains. Moreover, it allows the integration of additional image features and extended the functionality.

1.2.1   JAVA SETUP

First things first. LIRE requires that you have Java 1.6 (also called Java 6) or later installed, to work properly. Depending on your computer's operating system (OS) you may need to follow different instructions to download, install, and test the latest version of Java supported by the type and version of OS that is currently running on your machine. Please check your computer and OS documentation for specific details and instructions.

Besides installing Java SE (standard edition), you might also want to install the Java SDK (software development kit), which allows you not only to run but also to compile Java code. If you are new to Java it is recommended to take a look at the Java Tutorial,[13] which is a great free resource for learning to handle and write Java code.

1.2.2   DOWNLOADING, UNPACKING, AND RUNNING LIREDEMO

With Java working on your computer, the next step is to download and unpack *LireDemo*, the Java desktop demo application for LIRE. The most recent version of LireDemo is hosted on Google Code.[14] After downloading and unzipping the zip file, go to the directory containing the unzipped files and run LireDemo using the file `Liredemo.jar`. This can either be done by double clicking the file or by opening a shell and running `java -jar LireDemo.jar`. This should result in an application window similar to the one shown in Figure 1.5.

Let us explore LireDemo's GUI and associated options. The main tasks are accessible through a toolbar (containing icons and names), located just below the menu bar of the application, and containing the following options.

1. **Index.** Create new indexes and add images to existing ones.

2. **Search.** Perform a query-by-example (QBE) search on an existing image index.

3. **Browse.** Browse an existing index (e.g., for choosing an image to be used as an example in a subsequent search).

4. **Mosaic.** Run a sample image mosaic creation application.

---

[13]http://docs.oracle.com/javase/tutorial/index.html
[14]http://code.google.com/p/lire/

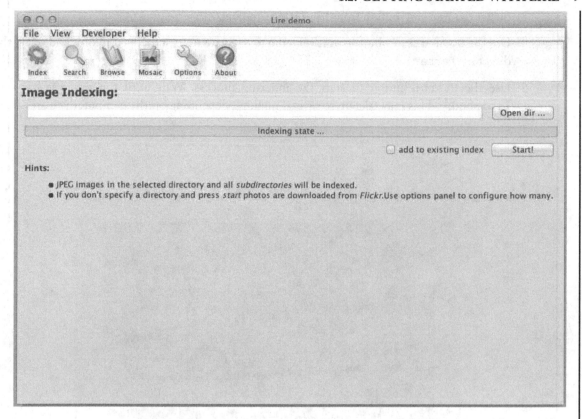

Figure 1.5: LireDemo: opening screen.

5. **Options.** Access the main options panel for configuring the number of results, the type of low-level features used, and a few other parameters.[15]

6. **About.** View administrative and copyright information.

In the remainder of this chapter we will explore some of LireDemo's main options separately.

1.2.3   INDEXING AN IMAGE COLLECTION

In order to index a set of images with LireDemo, follow these steps.

1. Download the "Ferrari" test data set from the book's companion website.[16]

2. Unzip the test data set to a directory (folder) of your choice.

---

[15]Please leave these options unchanged for now.
[16]http://lire-project.net/

3. Select the *Indexing* option from LireDemo's toolbar.

4. Use the `Open dir...` button of LireDemo to navigate to the unzipped photos and select the directory `ferrari`.

5. Use the `Start!` button to start the indexing process. Wait until the progress bar says `Finished`.[17] This may take up to several minutes, depending on the computer you are using.

Figure 1.6: Browsing the index.

1.2.4   BROWSING THE INDEX, SELECTING AN IMAGE, AND PERFORMING A SEARCH

Now it's time to use the recently built index to perform a search by visual similarity using an example image. Please follow these steps:

1. Click the "Browse" button to open the browse panel.

2. Scan through the indexed photos (using the up and down arrow buttons in the spinner in Figure 1.6) until you reach an image of interest, which will be used as an example for the search operation.

[17]The progress bar indicates approximately how much time remains and, at the end, reports the total time and the number of indexed files.

3. Once you reach an image you like, press the 'Search' button to trigger the search process.

4. After a few seconds, you should see a result screen (organized in rows of three images each), with the most relevant results at the top (Figure 1.7).

5. Take a closer look at the results list. You can right click on an image to open it in your operating system's default image viewer. Use double click on a search result to trigger another search with the clicked image as query image.[18]

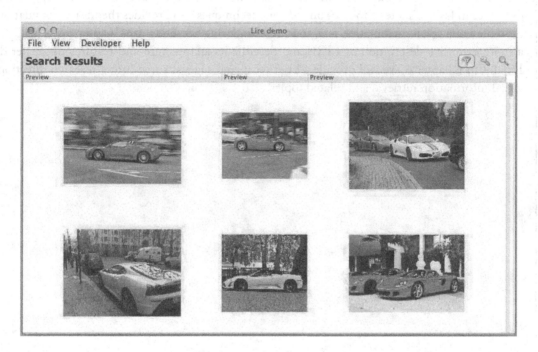

Figure 1.7: Top 6 search results using the image in Figure 1.6 (which appears as the best result—as expected—on the top-left) as an example.

Congratulations! At this point you have a working CBIR solution using LIRE. If you want to explore its functionality further (indexing other image folders, trying different visual features, etc.), go ahead and have fun! If you want to wait until you understand the concepts behind what you're doing, read on. There will be several LIRE- and LireDemo-based activities waiting for you in subsequent chapters.

---

[18]You might notice that, once you select an image, LireDemo paints an entire row of three images blue. Nothing to worry: once you double-click on the image, LireDemo knows which of the three images you have selected.

SUMMARY

In this chapter, we discussed the motivation for the design and implementation of visual information retrieval solutions, presented an overview of the main challenges involved in the process, and explained the steps involved in installing and interacting with a sample VIR solution based on LIRE.

The remainder of this book is structured as follows. In Chapter 2, we will present essential concepts and techniques from the field of information retrieval, illustrated by text-search examples using Lucene. In Chapter 3, we will focus on visual features and explain how they can be extracted from images and encoded in a way that is suitable for further processing, with implementation examples in Java. Chapter 4 focuses on indexing techniques and provides theoretical foundation and example code for representative text and visual indexing methods. In Chapter 5, we will present a more extensive description of LIRE and describe its architecture and features in greater detail. Finally, in Chapter 6, we will provide a brief recap and suggest some future research directions for visual information retrieval and related topics.

CHAPTER 2

Information Retrieval: Selected Concepts and Techniques

Visual information retrieval is well grounded in the field of *information retrieval*, which can be described as "the process of searching for (and retrieving) relevant information within a document collection." Information retrieval is a mature and well-established field of research. Visual information retrieval shares the same goals as text-based information retrieval, but focuses on documents containing visual information, i.e., images and videos.

The need for information retrieval is as old as the process of producing primitive documents. The first document collections date back thousands of years. The first (visual) documents currently known are the wall paintings of bison, horses, and other animals in the Cave of Altamira, Spain, estimated to having been produced more than 10,000 years ago. Writing started later and has a long history involving different alphabets and materials. Letters and pictograms were written on including silk, stone, clay, parchment, papyrus, and paper. The availability of written documents eventually led to the first document collections (archives and libraries). The largest and most significant library of the ancient world was the Royal Library of Alexandria founded some time between 305 BC and 283 BC. Along with the first libraries, the first library organization systems were born. Retrieval in those libraries of course was the work of librarians, who knew how to interpret organization systems, how to use catalogs and indexes and who knew where to find what.

With Gutenberg's printing press (invented around 1434–1440) and the spread of literacy, writing and documents reached general public. Library organization and retrieval went from card catalogs to micro films, and reached digital computers in the 1960s. *Information retrieval* itself was named as a concept in 1950. [59] In its first period, information retrieval was a purely academic topic. The second period of information retrieval started roughly in 1975, when commercial applications became available. The most significant milestone in the history of information retrieval was the availability of the World Wide Web, invented and introduced by Tim Berners-Lee in 1989. The new information age has allowed access to a vast wealth of information—more than ever before—and created a huge demand for a special type of practical application of information retrieval concepts, known simply as *(web) search engines*. In 1998, link analysis revolutionized information retrieval on the web with algorithms like PageRank and HITS implemented and used by Google and Teoma [42].

2.1   BASIC CONCEPTS AND DOCUMENT REPRESENTATION

Information retrieval is not about data. Information retrieval is about finding the information most relevant to a user's information need. According to [79] there is a clear difference between information and data retrieval. Two of the most dramatic differences are: (i) data retrieval needs a formal, artificial query language, e.g., SQL, whereas information retrieval deals with natural languages queries; and (ii) data retrieval returns *all* data matching the query exactly, while information retrieval returns just the best matches and will most likely return incomplete results.

Basically, information retrieval (Figure 2.1) deals with *documents*. The set of available documents is called document *corpus*. Users, who want to retrieve documents from a corpus, have a particular *information need*, like "I want to book a hotel in Atlanta next month and I need to find the phone number of a suitable hotel to call and reserve a room." This information need is abstract and not known to the retrieval system. The user just formulates a *query*, e.g., "hotel Atlanta." The retrieval system then accesses the *index*, a data structure for efficient retrieval, where all documents from the corpus are *indexed*. In information retrieval there is no perfect, or all-true result. So often we settle for "good enough" or "best under the circumstances." In other words, being "somewhat relevant" is better than nothing. This measure of relevance to the user is called *relevance function*. It decides numerically how relevant a document is for the user judging on the user query. The most relevant results are then presented to the user.

2.1.1   VECTOR RETRIEVAL MODEL

Retrieval models [6] provide a solid theoretical framework to design and evaluate new and innovative ideas and to test new approaches. The *vector retrieval model* is one of the most prominent ones. In the vector retrieval model, documents are points in a vector space. The terms of a corpus span the vector space and the vectors representing documents indicate which terms occur in which documents. Table 2.1 gives an example document term matrix of five documents. The first document, for instance, has a term vector of $d_1 = (2, 0, 1, 0, 0, 0, 0, 0, 2)$.

In the vector retrieval model, the similarity between documents d_i and d_j is typically determined by the *cosine coefficient* $s_c(d_i, d_j)$.

$$s_c(d_i, d_j) = \frac{d_i d_j}{|d_i||d_j|} .$$

Given our example the similarity of the first and second document $s_c(d_1, d_2)$ and between the second and the third document $s_c(d_2, d_3)$ can be determined by

$$s_c(d_1, d_2) = \frac{d_1 d_2}{|d_1||d_2|} = \frac{(2,0,1,0,0,0,0,0,2)(1,0,0,0,0,1,0,1,0)}{\sqrt{2^2+1^2+2^2}\sqrt{1^2+1^2+1^2}} = \frac{2}{3\sqrt{3}} \approx 0.385$$

$$s_c(d_2, d_3) = \frac{d_2 d_3}{|d_2||d_3|} = \frac{(1,0,0,0,0,1,0,1,0)(1,0,0,0,2,1,0,0,0)}{\sqrt{1^2+1^2+1^2}\sqrt{1^2+2^2+1^2}} = \frac{2}{\sqrt{3}\sqrt{6}} \approx 0.471 .$$

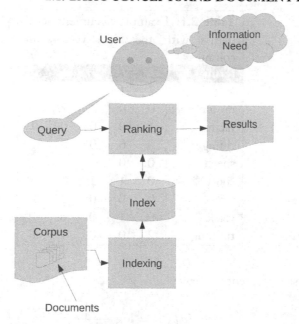

Figure 2.1: A general overview on the basic concepts of information retrieval.

Based on the cosine coefficient we can assume that the similarity between the second and the third document is higher than the similarity between the first and the second. As can be seen in Table 2.1, d_2 and d_3 have two terms in common $\{car, sport\}$, while d_1 and d_2 just have one single term in common $\{car\}$.

In the example at hand the number of occurrences of a term occurs in a document, known as *term frequency*, is taken into account. The raw term frequency $tf_r(t_i, d)$ of a term $t_i \in d$ in a document d is the actual number (count) of occurrence of a term. The normalized term frequency is then

$$tf(t_i, d) = \frac{tf_r(t_i, d)}{\max_{t \in d}(tf_r(t))} .$$

Term frequency is the first step towards *weighting schemes*. With weighting schemes, elements of the term vector are assigned numbers to reflect their actual importance for the document. The idea behind term frequency is that terms that have a higher frequency are more relevant for a document, i.e., those terms better describe the document's semantics. In the same example, words such as "car" have a high *document frequency*, which indicates the number of documents in which they appear. Since they occur in a large part of the documents in a corpus, they need to be assigned a lower weight to reflect their relatively lower discriminative power. A common and practical weighting

Table 2.1: Example document term matrix with the term vectors in columns

term	d_1	d_2	d_3	d_4	d_5
car	2	1	1	1	0
motorcycle	0	0	0	2	1
racing	1	0	0	0	0
riding	0	0	0	1	1
speed	0	0	2	0	1
sport	0	1	1	0	0
street	0	0	0	1	1
track	0	1	0	0	1
training	2	0	0	1	1

scheme is the inverse document frequency

$$idf(t_i) = \log \frac{|D|}{df(t_i)} \,,$$

where $df(t_i)$ is the number of documents $d \in D$ in the corpus D where t_i has a term frequency $tf(t_i, d) > 0$. Note at this point that the term frequency solely depends on a term's occurrence in a single document, but the inverse document frequency of a term is based on the term's occurrence in the whole corpus. Normalized term frequency, document frequency, and inverse document frequency for our example are given in Table 2.2.

Table 2.2: Example document term matrix with the term vectors in columns

Term	d_1	d_2	d_3	d_4	d_5	df	idf
car	1	1	0.5	0.5	0	3	0.097
motorcycle	0	0	0	1	1	2	0.398
racing	0.5	0	0	0	0	1	0.699
riding	0	0	0	0.5	1	2	0.398
speed	0	0	1	0	1	2	0.398
sport	0	1	0.5	0	0	2	0.398
street	0	0	0	0.5	1	2	0.398
track	0	1	0	0	1	2	0.398
training	1	0	0	0.5	1	3	0.222

The most prominent weighting scheme in text retrieval is the tf*idf scheme, where the weight $w_{i,j}$ of a term t_i in a document d_j is

$$w_{i,j} = tf(t_i, d_j) \cdot idf(t_i) = \frac{tf_r(t_i, d_j)}{\max_{t \in d_j}(tf_r(t))} \log \frac{|D|}{df(t_i)} \ .$$

Employing tf*idf in our example leads to new similarity values between the first and second document $s_{tfidf}(d_1, d_2)$ and between the second and the third document $s_{tfidf}(d_2, d_3)$

$$s_{tfidf}(d_1, d_2) = \frac{(0.097,0,0.349,0,0,0,0,0,0.222)(0.097,0,0,0,0,0.398,0,0.398,0)}{\sqrt{0.181}\sqrt{0.326}} \approx 0.029$$

$$s_{tfidf}(d_2, d_3) = \frac{(0.097,0,0,0,0,0.398,0,0.398,0)(0.048,0,0,0,0,0.398,0.199,0,0,0)}{\sqrt{0.326}\sqrt{0.2}} \approx 0.248$$

Comparing the results of the similarity based on the raw term frequency and the similarity based on tf*idf one can see that the ranking is not changed. The similarity between the second and the third document is still higher than the similarity between the first and the second. However, there is one significant change. Due to the fact that the term *car* has such a low inverse document frequency, the similarity between d_1 and d_2 is very close to 0, so they can not be considered similar in the way they had been without tf*idf. Also the relative difference between the two similarity values has changed. $s_{tfidf}(d_2, d_3)$ is roughly 8 times higher than $s_{tfidf}(d_1, d_2)$ compared to a factor of 1.22 in the example employing the raw term frequency.

For search scenarios, where users enter queries, a query is considered in the nearly the same way as a document. However, there is the slight difference that the query is not part of the corpus. Also, term frequency might not be too informative within a query, since people typically do not enter words more often than once in a typical query. A common scheme for query weighting is [6]

$$w_{i,q} = (0.5 + \frac{0.5 \cdot tf_r(t_i, q)}{\max_{t \in q}(tf_r(t))}) \log \frac{|D|}{df(t_i)}$$

so that for a query the term frequency is always above 0.5 and the inverse document frequency is taken from the document corpus. So a query like "motorcycle sport training" would result in our example in a vector $q = (0, 0.398, 0, 0, 0, 0.398, 0, 0, 0.222)$. Querying the corpus and ranking the similarity to the query in descending order gives the result list below (Table 2.3).

Table 2.3: Result list for our example

rank	d_i	$s_{tfidf}(d_i, q)$
1	d_5	0.586
2	d_4	0.550
3	d_2	0.511
4	d_3	0.366
5	d_1	0.285

2.2   RETRIEVAL EVALUATION

The main goal of information retrieval systems is to find the most relevant documents to satisfy the user's information need. In research, however, we need to compare different approaches to ensure that novel developments lead, indeed, to better retrieval results. For each and every retrieval algorithm we assume that it returns answers to a query as a ranked list of results, or simply an answer set $A \subseteq D$ being part of the corpus D. The best case is the one in which all documents of the set of relevant results $R \subseteq D$ are found. $R_a = R \cap A$ denotes the set of documents being found and relevant at the same time. Figure 2.2 uses a Venn diagram to put these quantities and concepts in context.

At a document level, depending on the document corpus, the subset of relevant documents, and the subset of found (retrieved) documents, each individual document d can be classified as:

- true positive (TP): if $d \in R_a$ (d is in the answer set *and* is relevant);

- true negative (TN): if $d \in D \setminus (R \cup A)$ (d is *not* in the answer set and is *not* relevant);

- false positive (FP): if $d \in A \setminus R$ (d is in the answer set but *not* relevant);

- false negative (FN) : if $d \in R \setminus A$ (d is *not* in the answer set but relevant).

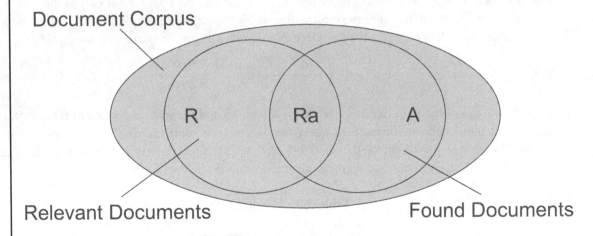

Figure 2.2: Venn diagram showing the typical relation between document corpus, documents relevant to a query, and actually retrieved answers.

The most prominent measures for the quality of a result set are: *precision p*, defined as the fraction of relevant documents in the retrieved result set; and *recall r*, defined as the fraction of relevant documents that were retrieved.

$$p = \frac{|R_a|}{|A|} = \frac{|R \cap A|}{|A|}$$

$$r = \frac{|R_a|}{|R|} = \frac{|R \cap A|}{|R|}$$

With an example result set $A = \{d_5, d_{20}, d_{27}, d_{31}, d_{40}, d_{54}, d_{63}, d_{73}, d_{78}, d_{98}\}$, the relevant set $R_a = \{d_{05}, d_{40}, d_{54}, d_{78}\}$ and a set R with $|R| = 12$ we have a precision $p = \frac{4}{10} = 0.4$ and a recall of $r = \frac{4}{12} = 0.333$. In our example d_5 would be a true positive and d_{20} would be a false positive.

Precision and recall are usually a trade-off, often represented in the form of P-R (Precision-Recall) graph [6] that shows that, usually, as the size of the result set increases, so does the recall, at the expense at lower precision. Since the two measures are tightly interconnected, it is often convenient to combine precision and recall in one single measure. A common combination is the *harmonic mean*, also called F_1 measure.

$$F_1 = 2 \cdot \frac{p \cdot r}{p + r}.$$

Precision, recall, and F_1 measure the quality of answer sets. Extending the purely set theoretic view would involve judging the ranking function to measure the quality of a result *list* instead of a result *set*. The easiest and most intuitive way to compute the quality of result lists is the *precision at k* measure. One basically cuts the result list after the first k elements and determines precision at that point. This especially makes sense in the context of the web, where users typically just view the first few results but algorithms retrieve thousands of results. In those cases, *precision at 10* then gives a rather good estimate of how accurate the first page of results is. For mobile applications, where the screen is smaller, *precision at 3* might be the measure of choice.

Testing a retrieval algorithm is typically an effort involving multiple different queries and their corresponding sets (or lists) of relevant documents. A pair $t_i = (q_i, R_i)$ combining a query q_i and the respective result set R_i is typically called a *topic*. Measures such as recall, precision, F_1, and precision at k are typically averaged over many topics to get a reliable measure for the quality of the retrieval algorithm. A combined measure for multiple topics $T = \{t_i | t_i = (q_i, R_i), q_i \in Q, R_i \subseteq D\}$ and the respective results as ranked lists a_i is the *mean average precision* (MAP).

$$MAP = |T|^{-1} \cdot \sum_{t_i \in T} ap(t_i)$$

$$ap(t_i) = |R_i|^{-1} \cdot \sum_{k=1}^{n} p_{at}(k) \cdot rel(k)$$

$$rel(k) = \begin{cases} 1, & \text{if there is a relevant document at position } k \\ 0, & \text{if there is no relevant document at position } k \end{cases}$$

with $ap(t_i)$ being the average precision, $p_{at}(k)$ being the precision at k for the result list in question, and $rel(k)$ gives 1 if for the result list in question a relevant documents is found at position k.

Example 2.1 Given a corpus D with 100 documents, identified by their number $D = \{1, 2, 3, ..., 100\}$, a topic t with the set of relevant results $R_t = \{3, 4, 8, 10, 12, 16, 30, 33, 34, 35, 42, 48, 57, 61, 62, 63, 87, 88\}$. A retrieval algorithm has returned the result list $l = (\mathbf{34}, 72, \mathbf{3}, \mathbf{8}, \mathbf{87}, 95, \mathbf{12}, 97)$.[1]

Precision, recall, and F_1 are then

$$p = \frac{5}{8} = 0.625, r = \frac{5}{18} \approx 0.278 \text{ and } F_1 = 2 \cdot \frac{0.625 \cdot 0.278}{0.625 + 0.278} \approx 0.385 .$$

Precision at k $p_{at}(k)$ is easy to determine. $p_{at}(1) = 1$, $p_{at}(3) = \frac{2}{3}$, and $p_{at}(5) = \frac{4}{5}$. $ap(t)$ is a bit more confusing, but still not too complicated. We need to look at the result list sequentially from the first position down to the last. For each relevant result we add up the precision at k, with k being the position currently looked at. The sum then is divided by the actual number of relevant documents:

$$ap(t) = \frac{1}{18} \cdot (\frac{1}{1} + \frac{2}{3} + \frac{3}{4} + \frac{4}{5} + \frac{5}{7}) \approx 0.218 .$$

While all the above measures work with a set of results with an arbitrary number of results, two more performance measures useful in cases when there is just one relevant result are worth mentioning, namely the *error rate* and the *inverted rank*. They are defined as follows.

The error rate *er* measures the fraction of topics $t \in T$, that did not find the proposed result at the top of the result list.

$$er = |T|^{-1} \sum_{t \in T} err(t)$$

$$\text{with } err(t) = \begin{cases} 0, & \text{if there is a relevant document at position 1} \\ 1, & \text{if there is no relevant document at position 1} \end{cases}$$

Note that the error rate is related to the precision at one $er = 1 - p_{at}(1)$.

The inverted rank $ir(t)$ of a topic t is $ir(t) = \frac{1}{n}$ if the first (or only) relevant result is found at rank n. The mean inverted rank *mir* is the average of the inverted rank of all topics:

$$mir = |T|^{-1} \sum_{t \in T} ir(t) .$$

---

[1]Note that for the sake of readability the result in *l* that are relevant are set in bold face.

2.3 TEXT INFORMATION RETRIEVAL WITH LUCENE

Lucene is (i) a prominent open source text retrieval engine written in Java and (ii) the name of a top level Apache project [55]. The Lucene project is organized in several sub projects, whereas *Lucene Core* is the historical starting point and the original *Lucene*, the plain text search engine. Several initiatives starting from Lucene have made to Apache top level projects, like *Hadoop*, the well known cloud computing solution, and *Nutch*, the fully fledged web indexing and retrieval solution.

At the time of writing this book, Lucene is available in version 4.0 and supports a wide range of professional features including clever index management strategies, scalable and fast indexing, a stable API and integration of many tools necessary for solid text retrieval. The basic data structure in Lucene is the *index*. It is accessed with implementations of the classes `IndexReader` for read access and search and `IndexWriter` for write access. Storage of indexes is managed by implementations of the `Directory` class, whereas different implementations allow for different storage media, like hard disks, main memory, network, or memory cached indexes.

Creating an index and indexing text documents can be done in a few lines of code. Listing 2.1 gives a simple example on how to index text documents with Lucene 4.0. First, an `IndexWriter` compatible to Lucene version 4.0 is created. The index is written to the directory "index" and uses the `SimpleAnalyzer` to analyze text in the indexing step (cp. lines 1-3). Note that there are multiple analyzer implementations available, which allow for tokenizing, stemming and removal of stop words. For each document a Lucene *Document* instance is created (cp. line 6). For each document we add an id and the actual document text. Lucene can either store the text or just retain the indexing data, which reduces size significantly (cp. line 7-8). Each document is added to the index (cp. line 9) and finally the index is closed (cp. line 11). To check if the code works and the index contains everything intended developers can use Luke,[2] an open-source index viewer for Lucene. Luke can show the documents and terms indexed, execute searches, evaluate different analyzer, etc.

Listing 2.1: Simple example for indexing text documents with Lucene 4.0

```
1   IndexWriterConfig conf = new IndexWriterConfig(
2          Version.LUCENE_40, new SimpleAnalyzer(Version.LUCENE_40));
3   IndexWriter iw = new IndexWriter(FSDirectory.open(new File("index")), conf);
4   for (int i = 0; i < document.length; i++) {
5      String docText = document[i];  Document d = new Document();
6      d.add(new StringField("id," Integer.toString(i), Field.Store.YES));
7      d.add(new TextField("text," docText, Field.Store.YES));
8      iw.addDocument(d);
9   }
10  iw.close();
```

For search with Lucene the class `IndexSearcher` is used. It needs an `IndexReader` to access the index and return a `TopDocs` object instance containing a ranked list of documents from a search. The queries are either created with classes implementing the class `Query` or parsed from a string using the `QueryParser`. Listing 2.2 shows a simple example used to search the index created in Listing 2.1. First the `IndexReader` is used to open the index and the `IndexSearcher` is created (cp.

[2]http://code.google.com/p/luke

lines 1-2). The `QueryParser` is created to be compatible to Lucene version 4.0 and the default field named "text" and the default analyzer are given (cp. lines 3-4). The actual search returns a TopDocs instance (cp. line 5). Then results are iterated and the documents' ids and their relevance scores are printed to `System.out` (cp. lines 6-9). Note that it is recommended to use the same analyzer for search and for indexing. Otherwise, search will yield unexpected results.

Listing 2.2: Simple example for searching text documents with Lucene 4.0

```
1  IndexReader indexReader = DirectoryReader.open(FSDirectory.open(new
2  File("index"))); IndexSearcher is = new IndexSearcher(indexReader);
3  QueryParser qp = new QueryParser(Version.LUCENE_40,
4      "text," new SimpleAnalyzer(Version.LUCENE_40));
5  TopDocs topDocs = is.search(qp.parse("query string"), 5);
6  for (int i = 0; i < topDocs.scoreDocs.length; i++) {
7    ScoreDoc scoreDoc = topDocs.scoreDocs[i];
8    System.out.println(scoreDoc.score + ": " + indexReader.document(
9        scoreDoc.doc).getValues("id")[0]); }
```

Advanced use of Lucene includes distributed indexes, custom relevance functions, stemmers, multi-field searches, etc. An extensive discussion of these and other advanced methods can be found in the book *Lucene in Action* [55].

SUMMARY

In this chapter, we introduced basic principles and concepts from information retrieval, which are necessary to understand the rest of the book. We further discussed the most basic retrieval evaluation measures, which also find their use in content based image retrieval. In a practical Subsection 2.3 we have given a short introduction to the text search engine Lucene and have shown how to use it to build your own search engine. Note that we just scratched the surface of information retrieval and there is much more to learn about other retrieval models, relevance functions, query and document processing, and retrieval evaluation. General information on information retrieval for further reading is for instance given in [6]. Challenges in this field are summarized in [4]. Other notable milestone not covered here but which we want to point out are latent semantic indexing (LSA) [18], the BM-25 weighting scheme [67] [68], and PageRank, the relevance function used by Google in their web search engine [11], [42].

PROBLEMS

1.1 Given a corpus D with 100 documents, identified by their number $D = \{1, 2, 3, ..., 100\}$, a topic t with the set of relevant results $R_t = \{3, 4, 8, 10, 12, 16, 30, 33, 34, 35, 42, 48, 57, 61, 62, 63, 87, 88\}$. A retrieval algorithm has returned the result list $l = (22, 16, 34, 1, 11, 42, 63, 18, 35, 72, 88)$. Find out precision, recall, F_1, precision at $k \in \{1, 3, 5\}$, and the average precision.

1.2 Download the latest version of Lucene[3] and create a Java project using it. Use the code from Listing 2.1 to create your own indexing routine and employ it to index several small documents. The documents can be collected for instance from Wikipedia[4] by copy/pasting articles or paragraphs from several pages. Then use Luke[5] to inspect your index and test some search queries. Finally, program a search routine for querying your index.

[3]http://lucene.apache.org/core/
[4]http://en.wikipedia.org
[5]http://code.google.com/p/luke

CHAPTER 3

Visual Features

Features that can be extracted from an images's raw pixel values are called *visual features*. Visual features can be broadly classified into two main groups: *global* features, which attempt to encode and describe global image properties (e.g., color and edge histograms), and *local* features, which focus on interest points within the image that carry enough information about the image as to enable further indexing and retrieval operations. We begin the chapter by presenting selected basic image processing concepts, followed by a technical discussion of the main feature extraction and representation techniques used in VIR, metrics, and distance functions used to evaluate different features, and tips on how to use LIRE to perform visual feature extraction.

3.1 DIGITAL IMAGING IN A NUTSHELL

Human vision is a remarkably complex process that starts with the eye and is fully accomplished by specialized regions in the brain (Figure 3.1). The basic goals of the eye are: (a) to capture light (electromagnetic radiation in the 400–700 nm wavelength range) reflected by objects in a scene; (b) to encode the information through electrochemical reactions; and (c) to transmit the corresponding electrical signals down the optic nerve to the brain, where it is interpreted, so that we can make sense of the scene, or simply, *see* what is in front of us.

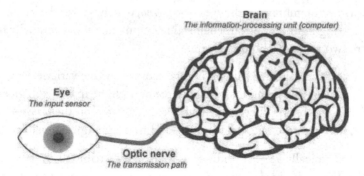

Figure 3.1: Simplified view of the connection from the eye to the brain via the optic nerve. Adapted and redrawn from [77].

Figure 3.2 shows a simplified cross section of the human eye. The eye's ability to focus on an image is provided by the *lens*, whose movements are controlled by specialized *ciliary muscles*. The anterior portion of the lens contains an *iris diaphragm*, which regulates the amount of light that enters the eye. The central opening of the iris is called *pupil* and varies its diameter—from 2–8 mm, approximately—in a way that is inversely proportional to the amount of incoming light.

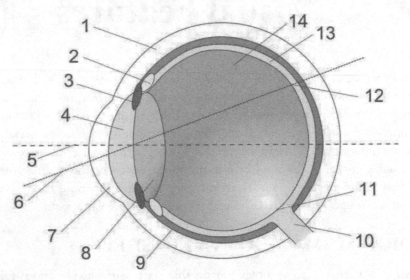

Figure 3.2: The eye: a cross-section view. 1 - sclera; 2 - ciliary body; 3 - iris; 4 - pupil, and anterior chamber filled with aqueous humor; 5 - optical axis; 6 - line of sight; 7 - cornea; 8 - crystalline lens; 9 - choroid; 10 - optic nerve; 11 - optic disc; 12 - fovea; 13 - retina; 14 - vitreous humor. Courtesy of Wikimedia Commons.

The innermost membrane of the eye is the *retina*, which is coated with discrete photoreceptors, capable of converting light into electrochemical reactions that will eventually be transmitted to the brain. There are two types of photoreceptors.

- *Cones*: Cones (typically 6–8 million in total) come in three varieties—S, M, and L, as in *short*, *medium*, and *long* (wavelengths), roughly meaning light in the red, green, and blue portions of the visible spectrum. They are primarily concentrated on the *fovea*—the central part of the retina, aligned with the main visual axis—and they are highly sensitive to color.

- *Rods*: Rods (typically 75–150 million in total) are distributed over the entire retinal surface (except for the *optic disc*, a region of the retina that corresponds to the perceptual *blind spot*). Rods are not sensitive to color, but are sensitive to low levels of illumination, and therefore responsible for dim-light (scotopic) vision.

Digital imaging in some way mimics the human eye. Light enters a digital camera through a lens (or multiple lenses) and is projected onto a sensor. The sensor usually has three types of photo sensitive elements (red, green, and blue) and converts the incoming light into electrical signals. The resulting analog signal is then converted to a digital representation, which consists of two main steps: *sampling* and *quantization*. Sampling involves selecting a finite number of points within each dimension of the image, whereas quantization implies assigning an amplitude value (within a finite range of possible values) to each of those points. The result of the digitization process is a *pixel array* (or *raster*), which is a rectangular matrix of picture elements whose values correspond to their intensity (for monochrome images) or color components (for color images). Contemporary cameras produce pixel arrays with more than 1,000 pixels per dimension, i.e., greater than 1 megapixel (MP) in terms of total number of pixels, with 256 colors per color channel (R, G, or B), resulting in a total of approximately 16 million colors (i.e., 256^3).

Figure 3.3 shows an example sampling and quantization process. The image on the left-hand side shows a circle with a continuous gradient of colors, from red (center) to dark red (edges). On the right-hand side, a sample for each cell of the raster has been taken and the color for a whole area is represented by a quantized color value.

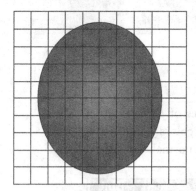 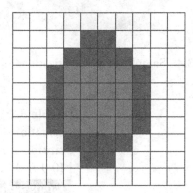

Figure 3.3: Sampling and quantization.

The effects of rigorous quantization and scaling can easily be seen in Figure 3.4. In the upper row the resolution of the image changes from 50×50, to 25×25, and 10×10 pixels. Visualizing the images at the same size allows us to see the pixelation (caused by too few samples) effects very clearly. In the lower row the number of colors of the 50×50 pixels images is varied from 256, to 4 and to 2 shades of gray. As the number of colors is decreased, the resulting images looks more artificial.

For the purposes of this book, a digital image is represented by a series of pixels, in a predefined order, where each pixel has a color value. We focus on pixels with RGB color, which means that the color of a pixel $p(x, y)$ at position (x, y) within a digital image is represented by a vector $p(x, y) = (r_{(x,y)}, g_{(x,y)}, b_{(x,y)})$. Table 3.1 gives some examples for common colors and their values in RGB.

Digital images are usually stored or transmitted in compressed format, to save disk space and transmission time. Compression methods can be *lossy*—when a tolerable degree of deterioration in the visual quality of the resulting image is acceptable—or *lossless*—when the image is encoded in its

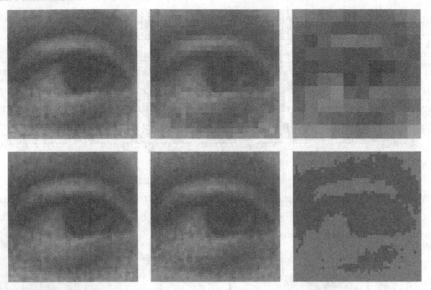

Figure 3.4: Sampling and quantization.

Table 3.1: Examples of RGB representation of selected colors with intensity for each channel normalized to the [0.0..1.0] range

Color name	Red	Green	Blue
red	1.0	0.0	0.0
green	0.0	1.0	0.0
blue	0.0	0.0	1.0
cyan	0.0	1.0	1.0
magenta	1.0	0.0	1.0
yellow	1.0	1.0	0.0
white	1.0	1.0	1.0
black	0.0	0.0	0.0

full quality [54]. The JPEG format is the most popular file format for photographic quality image representation. It is capable of high degrees of compression with minimal perceptual loss of quality. The Portable Network Graphics (PNG) is an increasingly popular file format that supports both indexed as well as true color images. Moreover, it provides a patent-free replacement for the GIF format and supports transparent pixels (where *transparency* is defined by a fourth color channel, the *alpha* channel). The compression of PNG depends heavily on the image content, therefore images with large uniform areas can be compressed to a very small size, while natural images, such as photos, result in rather big files (compared to their JPEG equivalent).

3.1.1 DIGITAL IMAGING IN JAVA

Java, particularly versions 6 and above, includes partial support for reading and writing PNG and JPG image files through the `javax.imageio` package. The class `javax.imageio.ImageIO` provides static convenience methods for reading and writing images. `ImageIO.read(...)` returns a `BufferedImage` object, which is the main representation of in-memory images in Java. To read and write pixels of an image one has the access the `WritableRaster` object within the `BufferedImage` with the `BufferedImage.getRaster()` method, which then offers setPixel(...) and getPixel(...) methods. Listing 3.1 gives sample code for opening an image and accessing its pixel values. The provided code prints out the RGB values for the first (upper left corner) pixel within the image. Pixel color values are within the [0, 255] range.

Listing 3.1: Opening an image in Java

```
1    String fileName = "sample_image.png;"
2    BufferedImage image = ImageIO.read(new FileInputStream(fileName));
3    WritableRaster raster = image.getRaster();
4    int[] tmpPixel = new int[3];
5    raster.getPixel(0, 0, tmpPixel);
6    System.out.println("p(0,0)=(" + tmpPixel[0] + ,'"" + tmpPixel[1]
7            + ,'"" + tmpPixel[2] + ")");
```

It is known that the Java implementation of Oracle does not support JPG files with color spaces other than RGB. In that case, one might need to employ third party software to decode some JPG files that cannot be read with the Java runtime. Robust JPEG decoders are integrated in packages such as ImageJ,[1] an image processing and analysis toolkit written in Java. Listing 3.2 shows the code necessary to load and decode an image with ImageJ and to convert it to a `BufferedImage` instance.

Listing 3.2: Opening an image with ImageJ

```
1    public static BufferedImage openImage(String path) {
2        ImagePlus imgPlus = new ImagePlus(path);
3        // converting the image to RGB
4        ImageConverter imageConverter = new ImageConverter(imgPlus);
5        imageConverter.convertToRGB();
6        // returning the BufferedImage instance
7        return imgPlus.getBufferedImage();
8    }
```

Manipulating pixel values (e.g., converting from color to grayscale) and saving the results back to disk is as easy as shown in listing 3.3. Note specifically in this listing that the `int[]` array containing the pixel's values is not created for every pixel, but before that in line 4. This saves time and memory as the same object can be re-used for every pixel of an image.

[1]http://rsbweb.nih.gov/ij/

Listing 3.3: Converting an image to gray scale

```
1    String fileName = "sample_image.png;''
2    BufferedImage image = ImageIO.read(new FileInputStream(fileName));
3    WritableRaster raster = image.getRaster(); int[] tmpPixel = new int[3];
4    int tmpValue = 0;
5    for (int x = 0; x < raster.getWidth(); x++) {
6        for (int y = 0; y < raster.getHeight(); y++) {
7            raster.getPixel(x, y, tmpPixel);
8            tmpValue = (int) (0.3*tmpPixel[0] + 0.6*tmpPixel[1] + 0.1*tmpPixel[2]);
9            tmpPixel[0] = tmpValue;
10           tmpPixel[1] = tmpValue;
11           tmpPixel[2] = tmpValue;
12           raster.setPixel(x, y, tmpPixel);
13       }
14   }
15   ImageIO.write(image, "png", new FileOutputStream(fileName));
```

3.2   GLOBAL FEATURES

The main goal of content-based image retrieval is to find *similar* images. While similarity itself is a concept that is hard to formalize, the problem is compounded by the need for comparing millions of images to a query image at search time—a very challenging task. A common approach consists in representing images using the minimal amount of information needed to encode its essential properties. This minimal information—called *image descriptor*—is usually extracted from the image's raw pixel values (and their coordinates)—a process known as *feature extraction*—and encoded in a numeric vector, the *feature vector*. Similarity, then, is defined by a suitable metric that computes a distance between two vectors. The design, extraction, and encoding of the image features as well as the choice of the metric makes up for a mathematical representation of visual similarity.

Example 3.1   Figure 3.5 shows two 4×4 pixels images, I_1 and I_2. The pixels in these images are either red, yellow, or blue. A simple image feature would consist of counting the colors in the images and assigning the pixel count to dimensions in a vector. Therefore, since there are 7 red pixels, 8 yellow pixels, and 1 blue pixel in the first image, $I_1 = (7, 8, 1)$. Respectively, the feature vector for the second image is $I_2 = (8, 4, 4)$. Knowing both feature vectors a distance function can be applied. A simple function would just sum the absolute differences between components in feature vectors, i.e., $|7 - 8| + |8 - 4| + |1 - 4| = 8$. Hence, 8 is the quantification of difference between the images I_1 and I_2. (Note that such a distance function is called a L_1 distance and will be explained in detail in Section 3.4.)

Ideally, descriptors should be: (i) *representative* of the contents of the image (or region) from which they were extracted; (ii) *robust* to image rotation, scaling, or translation (often referred to as *RST invariant* in the computer vision literature); and (iii) *compact*, since the number of dimensions and the range of possible values along each dimension are critical to search time behavior.

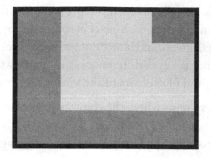 

Figure 3.5: Example images I_1 (left) and I_2 (right).

The naïve color-based descriptor derived in Example 3.1 is somewhat representative of the *global* contents of the image. One of its obvious limitations is its inability to convey the layout of the colors in each image. It is a rather compact descriptor: the feature vector size is 3, regardless of the size of the image. But what about robustness? A closer inspection shows that, despite its simplicity, it is robust against rotation by angles multiple of 90°. Scaling, however, leads to a different number of pixels in our example, and therefore to a different feature vector. In many cases, scaling operations also change the actual colors of pixels by approximating the color of merged or newly added pixels through their nearest neighbors. Our example feature is also robust against certain cases of translation, e.g., if pixels are switched in their positions, the actual feature vector is not changed.

3.2.1 COLOR FEATURES

"Color is one of the most obvious and pervasive qualities in our environment [26]." It is also a dominant feature in any content-based VIR system, due to the fact that the color information present in an image:

- can be computed in a relatively easy and straightforward way;
- is rather robust to background complications; and
- is mostly invariant to geometrical transformations, such as resize (scaling) or rotation.

Extracting color-based color features typically involves two main steps: (i) selection of a *color model* (or *color space*); and (ii) computation of a *descriptor* that encodes the color contents of an image—in a compact and discriminative way—according to the chosen color space.

Color Models

A *color model* (also called *color space* or *color system*) is a specification of a coordinate system and a subspace within that system where each color is represented by a single point. There have been many different color models proposed over the last 400 years [54]. Some of the most popular color models for VIR are: grayscale, RGB, HSV (and its variations), and HMMD. They are described below.

- **Grayscale:** The grayscale color model is actually an image representation through which the color information is removed and the intensity of each pixel is computed by: $Y = 0.3R + 0.6G + 0.1B$ (where R, G, and B are the red, green, and blue values of each pixel). The weights in this equation were chosen to model the human eye's differences in sensitivity to light stimuli in wavelengths that map to the red, green, and blue colors of the visible spectrum.

  Grayscale representations are used in VIR systems where the color information was not available in the first place (e.g., medical images, such as x-rays and CT scans). They can also be used by popular local visual descriptors, notably SIFT (Scale-Invariance Image Transform), where color doesn't play any role.

- **RGB:** The RGB color model is based on a Cartesian coordinate system, whose axes represent the three primary colors of light (R, G, and B), usually normalized to the range [0,1]. This information is usually available directly from the color raster for bitmap images, i.e., once an image file is read and decoded, no further conversions or calculations are needed to obtain its RGB representation.

  The number of discrete values of R, G, and B is a function of the *pixel depth*, defined as the number of bits used to represent each pixel: a typical value is 24 bits (3 color channels × 8 bits per channel). In VIR systems, the R, G, and B values are often *(re-)quantized* to a much smaller number of values, typically four quantization bins per primary color, resulting in a total of $4^3 = 64$ color combinations.

  The RGB color model is sometimes expressed in a normalized way, where the normalized colors are denoted r, g, and b and computed as: $r = \frac{R}{R+G+B}$, $g = \frac{G}{R+G+B}$, $b = \frac{B}{R+G+B}$, and $r + g + b = 1$.

  Another variant of the RGB color model is the opponent color space O_1, O_2, O_3, given by [78]:

$$
\begin{bmatrix} O_1 \\ O_2 \\ O_3 \end{bmatrix} = \begin{bmatrix} \frac{R-G}{\sqrt{2}} \\ \frac{R+G-2B}{\sqrt{6}} \\ \frac{R+G+B}{\sqrt{3}} \end{bmatrix}.
\tag{3.1}
$$

  In the opponent color space, the O_3 channel represents the intensity information, whereas O_1 and O_2 encode the color information.

  The RGB color space and its variants are used in a large number of color descriptors, e.g., color histogram, opponent histogram, rg histogram, RGB-SIFT, OpponentSIFT, C-SIFT, and rgSIFT, among others [78].

- **HSV:** The HSV model is part of a family of color models with the ability to dissociate the dimension of *intensity* (also called *brightness* or *value*) from the *chromaticity*—expressed as a combination of *hue* and *saturation*—of a color. Other, closely related, color models in this family are: HSI, HSB, and HSL.[2] The main advantages of the *HSV* color model (and its closely related alternatives) are its ability to match the human way of describing colors and to allow for independent control over hue, saturation, and intensity (value).

Conversion from RGB (often referred to as `sRGB` in Java documents) to HSV (referred to as `HSB`) in Java is quite straightforward due to the availability of methods `HSBtoRGB` and `RGBtoHSB` for objects of class `java.awt.Color`.

The HSV color space and its variants are used in several color descriptors, e.g., hue histogram and HSV-SIFT [78].

- **HMMD:** The HMMD (Hue-Max-Min-Diff) color space was developed in connection with MPEG-7 visual descriptors standardization efforts. It is based on the RGB and HSV models and was conceived specifically for CBIR.

The five components[3] of the HMMD color space are computed as follows:

$$H = \text{Hue component of the HSV color space} \tag{3.2}$$

$$Max = max(R, G, B) \tag{3.3}$$

$$Min = min(R, G, B) \tag{3.4}$$

$$Diff = Max - Min \tag{3.5}$$

$$Sum = \frac{Max + Min}{2} \tag{3.6}$$

- **Other color models:** there are *many* other color models in the image and video processing literature that were not included in this list. The interested reader may want to refer to Chapter 12 of [12] for additional information, including examples in Java.

Color histograms

Color histograms are the most intuitive and the most common visual descriptor. A color histogram is composed of bins each representing the relative amount of pixels of a certain color. Each pixel of an image is assigned to one color bin and increases the respective count. Intuitively, a color image with 16 M different colors would result in a histogram with 16 M bins. To limit the number of dimensions, the color information is typically *quantized*, so similar colors in the original color space

---

[2]Beware that there is no universal agreement in the literature regarding these variants: one paper's HSV may be the next paper's HSL.

[3]In spite of just four components being explicit in the name: Hue, Maximum, Minimum, Difference, the HMMD color space also includes a fifth one: Sum.

are considered as if they were identical and their frequency of occurrence is computed for the same bin. Color quantization is a critical step in the construction of a feature. The most straightforward color quantization approach consists in dividing the RGB color space into equal-sized partitions, like stacked boxes in 3D space, which has been shown to work rather well in [19].

Listing 3.4 gives an example of color quantization in RGB color space. The quantized colors correspond to the definition of 64-color RGB.

Listing 3.4: Quantization in RGB color space to 64 bins

```
1   int[] histogram = new int[64];
2   for (int i = 0; i < histogram.length; i++) histogram[i] = 0;
3   BufferedImage img = ImageIO.read(new FileInputStream("testImage.jpg"));
4   WritableRaster raster = img.getRaster();
5   int[] px = new int[3];
6   for (int x = 0; x < raster.getWidth(); x++) {
7       for (int y = 0; y < raster.getHeight(); y++) {
8           raster.getPixel(x, y, px);
9           int pos = (int) Math.round((double) px[2] / 85d) +
10                    (int) Math.round((double) px[1] / 85d) * 4 +
11                    (int) Math.round((double) px[0] / 85d) * 4 * 4;
12          histogram[pos]++;
13      }
14  }
```

In the HSV color space, however, linear quantization is a harder task, for two main reasons.

• The hue is represented by a color wheel and encoded as an angle (from 0°–360°) relative to an arbitrary (usually the *red* hue) reference. Therefore, even though the different colors can be easily divided into bins of a certain angle interval, there is an inherent discontinuity around 0°.

• If saturation is too low, i.e., $S = 0$, then no color at all is visible, and the value of Hue (H) is undefined. In those cases, the value (V) gives the shade of gray from black to white. If saturation is greater than 0, however, V defines how much white is in the color, i.e., how desaturated it is.

Even though the code given in Listing 3.4 effectively computes a color histogram, the resulting histogram cannot be used as a feature vector yet because it relies on absolute pixel counts, which are dependent on the image size. Resized pictures would have very different feature vectors, so the feature is not robust to scaling. Therefore, the histogram has to be normalized, so that the length of the feature vector is equal to 1 (see also Section 3.4). For example, for a histogram $H = (h_1, h_2, \ldots, h_k)$ with k components, the normalized histogram H^n, which serves well as a feature vector, can be (under the assumption that the length of a vector is the sum of its components)

$$H^n = (\frac{h_1}{\sum_{i=1}^{k} h_i}, \frac{h_2}{\sum_{i=1}^{k} h_i}, \ldots, \frac{h_k}{\sum_{i=1}^{k} h_i}) \, .$$

In summary, a descriptor based on color histograms requires that several choices be made, in the following order.

1. **Select a color space.** Image pixels are typically stored in RGB color. For all other color spaces, a color space conversion routine[4]—which takes up extra extraction time and memory—is needed.

2. **Define histogram length and quantize color.** The histogram length defines the size of the descriptor. So a feature vector based on a 1024-bin histogram takes up more storage space and requires more time for distance computation than a feature vector based on a 64-bin histogram. It is a design decision to find a tradeoff between descriptor size and effectiveness in image retrieval tasks.

3. **Normalize and quantize the histogram.** Normalization is necessary for achieving robustness against scaling. Additional strategies for normalization are presented in Section 3.4. After normalization, the resulting histogram values should be within the [0, 1] range. With a quantization function $q_k : V \rightarrow Q_k$ one can map values from $V = \{v|0 \leq v \leq 1\}$ to an arbitrary set of numbers $Q_k = \{0, 1, 2, 3, ..., k\}$ with $k \in \mathbb{N}$. So if histogram values h_i are quantized for instance with $k = 255$, i.e., $q_{255}(h_i) = (h_i^{q255})$, each bin of the histogram can be stored as a `byte` value, which would take one fourth of the space that would be required if each bin were stored as a `float` value, i.e., taking up 32 bits (4 bytes). While the actual effect of the quantization step on the retrieval performance is an issue better left to retrieval evaluation (Section 3.5), it has been shown in several scenarios that this step does not reduce precision and/or recall [14].

Fuzzy color

Color quantization is—in some use cases—unable to identify similar colors. Figure 3.6 (left-hand side) illustrates the problem. Two very similar shades of orange (look inside the two circles) correspond to hue right next to the border for quantization. The problem is that while these two shades of orange look very similar, they are quantized in such a way that one is assigned to *yellow* and the other one to *red*.

Fuzzy color quantization tries to tackle exactly this problem. Instead of discrete classification in either red or yellow, a membership function defines the degree of membership of a certain color to a color class. In our example, orange would be 50% yellow and 50% red. Figure 3.6 (right-hand side) gives an illustration of fuzzy color quantization. The dotted line gives the membership function for red, while the solid line gives the membership function for yellow. Note that in fuzzy set theory the sum of memberships of a single element (pixel) is always 1. So the membership functions always result in a degree of membership within the [0, 1] range. In the example given in Figure 3.6 the given two shades of orange would lead to 0.75 yellow and 0.25 red for the left sample and 0.25 yellow and 0.75 red for the color sample on the right. The histogram itself is then the sum of membership function values for all pixels per bin (i.e., color class).

---

[4]See http://www.f4.fhtw-berlin.de/~barthel/ImageJ/ColorInspector/HTMLHelp/farbraumJava.htm for color space conversion code in Java.

Figure 3.6: Crisp color quantization vs. fuzzy color quantization.

A detailed explanation of a specific fuzzy color quantization scheme for the *Color and Edge Directivity Descriptor* (CEDD) feature is given in [14]. In CEDD a 24-bin fuzzy color histogram is created based on the HSV color model. CEDD is also implemented in LIRE.

Dominant color

A minimal feature in terms of size can be constructed based on the most dominant color of an image. The approach is more or less the same as for the color histogram. After quantization a histogram is created and the bin with the highest peak is identified as dominant color. For comparison of features, the distance between dominant colors of images is computed. The expressiveness of a single dominant color is, of course, limited. Figure 3.7 shows four different images. The pairs A and B as well as C and D have the same dominant color and therefore a minimal distance. However, images B and C look very similar although their dominant color is different.

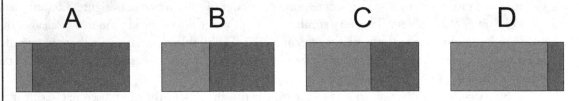

Figure 3.7: Examples to illustrate the problem of dominant color features.

In MPEG-7 a dominant color descriptor based on the three most prominent colors is proposed. While this compensates the problem posed by images with multiple dominant colors, distance metrics prove complicated and lead to a minimization problem, i.e., what is the minimal distance between two weighted sets of colors.

3.2.2   TEXTURE FEATURES

Texture is a powerful discriminating feature, present almost everywhere in nature. Texture similarity, however, is more complex than color similarity. Two images can be considered to have similar texture when they show similar spatial arrangements of colors (or gray levels), but not necessarily the same colors (or gray levels).

There are several possible approaches to represent and extract the texture properties of an image. Different authors use different classifications. According to Gonzalez and Woods [27], there are three main categories of texture models.

1. **Statistical models**. Statistical approaches to describe texture properties usually fall within one of these categories.

   • The use of statistical moments of the gray-level histogram of an image or region to describe its texture properties. The second moment (the variance) is of particular importance because it measures gray-level contrast and can therefore be used to calculate descriptors of relative smoothness. Histogram information can also be used to provide additional texture measures, such as *uniformity* and *average entropy*. Similarly, to what was said for color histograms as color descriptors, the main limitation of using histogram-based texture descriptors is the lack of positional information.

   • The use of descriptors (energy, entropy, contrast, homogeneity, etc.) derived from the image's gray-level co-occurrence matrix, originally proposed by [32].

2. **Spectral models**. These models rely on the analysis of the power spectral density function in the frequency domain. Coefficients of a 2-D transform (e.g., the Wavelet transform [13, 28, 41]) may be considered to indicate the correlation of a brightness pattern in the image. Coarse textures will have spectral energy concentrated at low spatial frequencies, while fine textures will have larger concentrations at high spatial frequencies.

3. **Structural models**. This category includes methods that describe textures in terms of primitive *texels* [71] in some regular or repeated relationship. This approach is appealing for artificial, regular patterns.

Texture is a broad concept, involving aggregations that often depend on data, context, and culture. Moreover, it is fundamentally a problem of scale, and that is why it is so difficult to find texture descriptors that work well in unconstrained databases.

Practically speaking, different texture descriptors have been proposed and have been used for various visual information retrieval tasks. In the following, we give an overview on two selected texture features: the Tamura features and the MPEG-7 Edge Histogram.

Tamura Features

A classical set of simple features are the so called *Tamura features* as introduced in [75]. Basically, six different characteristic features of a texture have been identified: (i) coarseness (coarse vs. fine); (ii) contrast (high contrast vs. low contrast); (iii) directionality (directional vs. nondirectional); (iv) line-likeness (line-like vs. blob-like); (v) regularity (regular vs. irregular); and (vi) roughness (rough vs. smooth).

The original publication [75] found that the first three proposed features correlate well with human perception. Note that the Tamura features describe texture on a global level, so their usefulness for pictures with complex scenes involving different textures is limited.

Edge histogram

The edge histogram feature is part of the MPEG-7 standard, a multimedia meta data definition including several features for visual information retrieval [72]. The edge histogram feature captures the spatial distribution of (undirected) edges within an image. First, the image is partitioned into 16 equal-sized, non overlapping blocks. For each block edge information is then processed and put into a 5-bin histogram counting edges in of the categories: vertical, horizontal, 35°, 135°, and non directional. This feature is mostly robust against scaling.

3.2.3 COMBINING COLOR AND TEXTURE

Intuitively, the next question would be how to combine the best of both worlds, i.e., color and texture. A simple example, as proposed and employed in [19], of such a feature is a scaled down instance of the image. Due to the scaling operation a whole sub-image is reduced to a single pixel. With an additional quantization step the number of colors is further reduced. The feature is then created from a histogram encoding for each bin the color of the responding sub image (or pixel in the scaled and quantized version). So if the feature vector $f = (0, 13, 63, 2, \ldots)$, we can assume that the upper left sub image has the dominant color 0, the second pixel from the left has color 13, the third one color 63, and so on. Figure 3.8 illustrates the steps of scaling and quantization.

Figure 3.8: Simplified color layout feature, from left to right: the original image, the scaled version, the scaled version quantized to the 256 color system palette.

This feature can be considered a hybrid one as it combines the color information with the actual spatial information of its occurrence. So there is implicit texture information hidden in there as well. A feature along these lines is the MPEG-7 color layout descriptor [72], which captures dominant color information in equally sized sub-images.

The general notion of features combining color and texture information is called *joint features*, as defined in [62]. Joint features can be implemented by encoding the color information into a histogram and using texture information to further divide the color bins. A simple example of such a combination is a color histogram with an additional dimension for the pixel rank. The rank of pixel $p_{x,y}$ is defined as the number of pixels in the local neighborhood whose intensity is less than the intensity of the pixel $p_{x,y}$. The result is a two-dimensional histogram that counts pixels based on color and their rank. Listing 3.5 gives sample code for extracting a joint histogram of the RGB 64 color space and pixel rank. Note that in this example for the sake of simplicity and extraction speed the border pixels at the edge of the image are omitted. Rank is computed based on a neighborhood determined by Manhattan distance $d_M = 1$, i.e., the *8-neighborhood* of the reference pixel.

Listing 3.5: Example code for extracting a joint histogram of RGB64 and pixel rank

```
1   int[][] histogram = new int[64][9];
2   for (int i = 0; i < histogram.length; i++) {
3       for (int j = 0; j < histogram[i].length; j++)
4           histogram[i][j] = 0;
5   }
6   BufferedImage img = ImageIO.read(new FileInputStream("testImage.jpg"));
7   WritableRaster raster = img.getRaster();
8   int[] px = new int[3];
9   for (int x = 1; x < raster.getWidth()−1; x++) {
10      for (int y = 1; y < raster.getHeight()−1; y++) {
11          raster.getPixel(x, y, px);
12          int colorPos = (int) Math.round((double) px[2] / 85d) +
13                         (int) Math.round((double) px[1] / 85d) * 4 +
14                         (int) Math.round((double) px[0] / 85d) * 4 * 4;
15          int rank = 0;
16          double intensity = getIntensity(px);
17          if (getIntensity(raster.getPixel(x−1, y−1, px))>intensity) rank++;
18          if (getIntensity(raster.getPixel(x , y−1, px))>intensity) rank++;
19          if (getIntensity(raster.getPixel(x+1, y−1, px))>intensity) rank++;
20          if (getIntensity(raster.getPixel(x−1, y+1, px))>intensity) rank++;
21          if (getIntensity(raster.getPixel(x , y+1, px))>intensity) rank++;
22          if (getIntensity(raster.getPixel(x+1, y+1, px))>intensity) rank++;
23          if (getIntensity(raster.getPixel(x−1, y , px))>intensity) rank++;
24          if (getIntensity(raster.getPixel(x+1, y , px))>intensity) rank++;
25          histogram[colorPos][rank]++;
26      }
27  }
28
29  private double getIntensity(int[] px) {
30      return (0.3*px[0] + 0.6*px[1] + 0.1*px[2]);
31  }
```

A similar approach can be found in the extraction of the *color correlogram* [35] where, instead of joining color with the pixel rank, the second dimension is based on the neighboring colors. Basically, each cell $h_{k,l}$ of the histogram gives an indication of the probability to encounter color c_l in the neighborhood of color c_k. A faster and more efficient version of the color correlogram is the auto color correlogram, which just counts how often a color finds itself in its immediate neighborhood. Figure 3.9 shows an extreme example where the color correlogram clearly outperforms the common color histogram. Using a color histogram would lead to the assumption that these two images are equal. Color correlogram features, however, would be significantly different. More details on color correlograms as well as implementation details are given in [35].

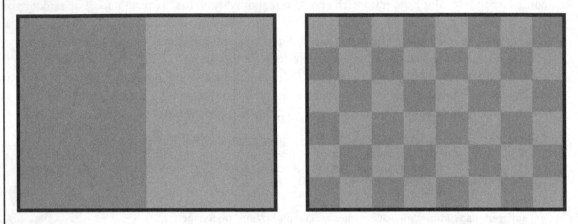

Figure 3.9: Examples to illustrate the benefit of the color correlogram feature.

A well-performing feature that combines fuzzy color and edge information is the *color and edge directivity descriptor* (CEDD) [14]. It combines a 24-bin fuzzy color histogram {black, gray, white, dark red, red, light red, dark orange, orange, light orange, dark yellow, yellow, light yellow, dark green, green, light green, dark cyan, cyan, light cyan, dark blue, blue, light blue, dark magenta, magenta, light magenta} with 6 different types of edges {no edge, non directional edge, horizontal edge, vertical edge, 45° edge, 135° edge}. After extraction, the 144 bins are normalized and the bins are further quantized to 3 bits. This results in an overall descriptor length of $144 \cdot 3 = 432$ bits, or 54 bytes. Compared to a RGB histogram of doubles (64 bit each) with 64 bins, which has a length of 512 bytes, CEDD is rather compact.

The *fuzzy color and texture histogram* (FCTH) [15] employs the same fuzzy color scheme but uses a more extensive edge description with 8 bins resulting in an overall feature with 192 bins. Here, too, the bins are quantized to three bins resulting in a very compact feature representation. A combination of CEDD and FCTH is the joint composite descriptor (JCD) [16].

3.3   LOCAL FEATURES

A local feature is "an image pattern which differs from its immediate neighborhood" [76]. It is usually associated with a change of one or more image properties, such as intensity, color, and texture. Local features can be interest points, corners, edges, or salient spots. Typically, some measurements are taken from a small (usually between 3×3 and 60×60 pixels) region (also known as *patch*) centered on a local feature and converted into descriptors. The descriptors can then be used for various applications, e.g., object recognition, texture recognition, building panoramas, recognition of object categories, and image retrieval [57].

At their lowest level, local feature extraction algorithms can be classified in three main categories: (i) *Corner detectors*, which includes the Harris detector and its variants (Harris-Laplace and Harris-Affine) and the SUSAN detector; (ii) *Blob detectors*, which includes the Hessian detector and its variants (Hessian-Laplace and Hessian-Affine), salient region detectors, as well as the popular DoG (difference-of-Gaussians) and SURF (speeded-up robust features) methods; and (iii) *Region detectors*, which include Maximally Stable Extremal Region (MSER) and segmentation-based methods (e.g., superpixels) [76].

Most contemporary local descriptors (e.g., Harris-Laplace, Hessian-Laplace, DoG, SURF) are robust to scaling and rotation and some are also robust to affine transformations, e.g., Harris-Affine, Hessian-Affine, and MSER [76]. The latter group is often referred to as "Affine Region Detectors" in the computer vision literature and have been the subject of a large number of recent studies (e.g., [58]).

From the perspective of image retrieval, the ideal local descriptor should be compact, fast to compute, and robust to scale and rotation. Two local descriptors achieved great popularity in the context of VIR systems during the past few years, SIFT and SURF.

3.3.1   SCALE-INVARIANT FEATURE TRANSFORM (SIFT)

The Scale-Invariant Feature Transform (SIFT)—originally proposed by Lowe [48]—combines a scale-invariant region detector and a descriptor based on the gradient distribution in the detected regions [57]. The descriptor is represented by a 3D histogram of gradient locations and orientations which identifies the regions' appearance compactly and robustly.

To compute the SIFT descriptor, we perform the following steps.

(i) Compute gradient orientation and amplitude maps over a 16×16 pixel region (i.e., patch) around the interest point.

(ii) Quantize the resulting orientation into eight bins spread over the range 0–360°.

(iii) Divide the 16×16 detector region into a regular grid of non overlapping 4×4 cells.

(iv) Compute an eight-dimensional histogram of the image orientations within each of these cells. (Each contribution to the histogram is weighted by the associated gradient amplitude and by distance so that positions further from the interest point contribute less.)

(v) Concatenate the $4 \times 4 = 16$ histograms into a 128×1 single vector, which is then normalized [65].

Figure 3.10 shows the main steps in computing the SIFT descriptor for a test image. Note that the first step is the color-to-grayscale conversion, since the SIFT descriptor *does not* take color information into account.

Figure 3.10: SIFT descriptor: (a) input image; (b) interest points (in yellow) overlaid on grayscale equivalent of input image; (c) grids of non overlapping 4×4 cells (in green) around each interest point; (d) a closer look at a cell near the top-left corner of the image.

3.3.2   SPEEDED-UP ROBUST FEATURES (SURF)

SURF is a performance-oriented scale and rotation-invariant interest point detector and descriptor. Its performance with respect to repeatability, distinctiveness, and robustness is comparable to previously proposed schemes, yet it can be computed and compared much faster [7]. Figure 3.11(a) shows an example of SURF feature extraction for the same input image used in Figure 3.10. The 10 strongest points have been selected. A quick comparison with the results obtained earlier using the SIFT descriptor (repeated in Figure 3.11(b) for convenient side-by-side comparison) shows that only a few interest points detected by SIFT appear among the strongest 10 points computed using SURF.

(a)   (b)

Figure 3.11: SURF descriptor: (a) 10 strongest points (in yellow) overlaid on grayscale equivalent of input image from Figure 3.10(a); (b) results using SIFT (Figure 3.10(b), repeated here for convenient side-by-side comparison).

3.4   METRICS, NORMALIZATION, AND DISTANCE FUNCTIONS

Features and their representations (i.e., descriptors) must be compared to each other to determine the visual similarity (i.e., closeness) between digital images. This is typically done using a *metric*. A *metric* (also called *distance function* or simply *distance*) is a function $d : X \times X \rightarrow \mathbb{R}$ for $x, y, z \in X$, which satisfies the following properties:

1. non negativity: $d(x, y) \geq 0$;

2. identity: $d(x, x) = 0$;

3. symmetry: $d(x, y) = d(y, x)$;

4. triangle inequality: $d(x, z) \leq d(x, y) + d(y, z)$.

Remember that features are typically normalized in a way that bin values, $\bar{h}_i$, of a normalized histogram, $\bar{H}$, represent the relative amount of image pixels belonging to the bin instead of the absolute count. Normalization also provides a certain robustness for scaling, as bin values are always in [0, 1] regardless of the size of the image. A histogram can easily be normalized by different means (depending on the metric that applies), an easy and efficient way is to divide histogram values of the original histogram H by the sum of its elements, or in other words, to normalize the length of the vector according to L_1 norm:

$$\bar{H} = (\frac{h_i}{\sum_j h_j}) \text{ with } \|\bar{H}\|_1 = 1 .$$

Other possible normalization functions are to normalize the length of the vector H based on the max norm:

$$\bar{H} = (\frac{h_i}{max_j(h_j)}) \text{ with } \|\bar{H}\|_\infty = 1$$

and based on the L_2 norm:

$$\bar{H} = (\frac{h_i}{\sqrt{\sum_j h_j^2}}) \text{ with } \|\bar{H}\|_2 = 1 .$$

For normalized feature vectors (i.e., histograms), certain metrics can be employed to compute the *visual distance* between pairs of images, meaning that images with small distance between them are more similar than image pairs with a large distance function value. Typical distance functions include:

$$
\begin{array}{lll}
\text{Manhattan or } L_1 \text{ distance} & d_{L_1}(x, y) & = & \sum_i |x_i - y_i| \\
\text{Euclidean or } L_2 \text{ distance} & d_{L_2}(x, y) & = & \sqrt{\sum_i (x_i - y_i)^2} \\
\text{Cosine coefficient} & d_{cc}(x, y) & = & \frac{x \cdot y}{|x| \cdot |y|} \\
\text{Jensen-Shannon divergence} & d_{jsd}(x, y) & = & \sum_i x_i \log \frac{2x_i}{x_i + y_i} - y_i \log \frac{2y_i}{x_i + y_i}
\end{array}
$$

L_1 is a popular metric, as it is rather fast to compute and performs well. L_2 is a classical approach, but all the multiplications and square roots tend to slow down the distance computation for large data sets. The cosine coefficient measures the angle difference between two vectors and is extensively used in text retrieval. The Jensen-Shannon divergence (JSD) has been shown to perform well with RGB histograms [19]. Other functions employed for image retrieval are the Earth Movers Distance (EMD) [69] and the Tanimoto coefficient, which has been proposed for CEDD and FCTH [14], [15]. An in-depth discussion of 38 different functions to determine the distance between feature vectors is presented in [22].

Example 3.2   In this example we revisit the naïve color descriptor for the image in Figure 3.5 (Example 3.1) which resulted in two 3-dimensional feature vectors ($I_1 = (7, 8, 1)$ and $I_2 = (8, 4, 4)$) and use these vectors to compute distances between the images, as follows:

$$
\begin{array}{lll}
L_1(I_1, I_2) & = & |7 - 8| + |8 - 4| + |1 - 4| & = & 8 \\
L_2(I_1, I_2) & = & \sqrt{(7 - 8)^2 + (8 - 4)^2 + (1 - 4)^2} & \approx & 5.099 \\
cc(I_1, I_2) & = & \frac{7 \cdot 8 + 8 \cdot 4 + 1 \cdot 4}{\sqrt{7^2 + 8^2 + 1^2}\sqrt{8^2 + 4^2 + 4^2}} & \approx & 0.879 .
\end{array}
$$

The results from L_1 and L_2 computations are just numbers, which would only make complete sense if compared with distances between other pairs of images in a dataset. The cosine coefficient, however, can be interpreted as a "coefficient of similarity" within the [0,1] range, which means that the two images are rather similar if compared using this simple three-bin color descriptor.

3.5   EVALUATION OF VISUAL FEATURES

Besides heuristic evaluation—which consists of experts considering the quality of the retrieval approach by testing the retrieval engine subjectively with arbitrary queries—quantitative evaluation of VIR solutions is crucial to document progress on development, to select appropriate pre-processing methods, features and metrics, and to compare different approaches. In order to evaluate quantitatively the performance of a VIR system one needs to adopt the appropriate figures of merit and test the solution against a suitable dataset. To compete against other VIR systems under the same set of circumstances, one may want to consider signing up for a challenge in the field. These three aspects—figures of merit, datasets, and challenges—are described next.

3.5.1   FIGURES OF MERIT

VIR systems use success measures derived from text retrieval, such as precision, mean average precision, precision at X, recall, and F_1 (Section 2.2). Typically, the developers decide upon which evaluation measure to use based on the scenario the VIR system is applied. If, for instance, only three results can be shown on a mobile device screen, precision at three is a good choice for evaluation. If there is just one relevant result per query, error rate or inverted rank are good choices.

There are figures of merit created especially for the purpose of VIR evaluation, e.g., the *average normalized modified retrieval rank* (ANMRR) [53], which was defined in the context of MPEG-7 color core experiments, and whose the goal was to give just *one* number indicating the retrieval quality over all queries.

3.5.2   DATASETS

During the early days of VIR, it was not uncommon to read papers in which new VIR methods were being tested using proprietary datasets, often a subset of what became known as the "Corel data set" (a collection of CDs with stock images organized in folders based on the categories to which they belonged). Over the past decade, a trend towards standardization of test datasets can be recognized. There are many datasets suitable for VIR evaluation that are now publicly available online. One of the most prominent and widely used test datasets is the SIMPLIcity dataset [81]. This dataset features 1,000 different images of size 256×384 or 384×256 pixels showing scenes from 10 different categories. The pictures are numbered and the numbers 0–99 show the first category, 100–199 the second category, and so on. It is assumed that if you search for an image out of category x all other images of category x are relevant results. All images that are not in category x are not considered relevant. Therefore, the SIMPLIcity data set contains 1,000 topics with a set of 99 relevant results each (not counting the query image in). Figure 3.12 gives examples for all of the ten categories: tribal, beach, buildings, buses, dinosaurs,[5] elephants, flowers, horses, mountains, and food.

Figure 3.12: Examples for each of the ten categories from the SIMPLIcity dataset [46, 81].

For a good overview of how different low-level features (and associated metrics) can be compared against several publicly available datasets (including SIMPLIcity) using two main figures of merit (error rate and mean average precision—MAP), we recommend the article by Deselaers et al. [19].

The SIMPLIcity data set has been used extensively over the years. However, with the advent of cheap and powerful computer hardware, larger datasets have been published. One of the most prominent current datasets is the MIRFLICKR image collection [36, 37]. The dataset features photos and metadata—including user annotation, tags, and EXIF data—collected from the Flickr website. All photos in the dataset are available under the Creative Commons license. The first release in 2008 contained 25,000 photos, the second release in 2010 features 1,000,000 photos.

Other notable datasets for VIR evaluation include: (i) the "Object and Concept Recognition for Content-Based Image Retrieval" image database—University of Washington;[6] (ii) the "Holidays" dataset—INRIA;[7] and (iii) UCID: An Uncompressed Color Image Database—Nottingham Trent University, U.K.[8]

3.5.3   CHALLENGES

During the past decade, the multimedia community has developed several prominent benchmarking initiatives, which have helped significantly to advance the state of the art in several fields, including visual information retrieval. A benchmark task consists of: a problem definition, a data collection, ground truth, and an evaluation metric [43]. Benchmarking efforts are often formatted as a challenge

---

[5]As the attentive reader might already have guessed, the dinosaur category is not based on original photos. All images in this category are artificial.
[6]http://www.cs.washington.edu/research/imagedatabase/
[7]http://lear.inrialpes.fr/~jegou/data.php
[8]http://homepages.lboro.ac.uk/~cogs/datasets/ucid/ucid.html

open to the research community at large, for which several leading universities and research labs submit their best runs and compete for the top spot in different categories.

The most relevant contemporary challenge in the context of VIR is ImageCLEF,[9] established in 2003. Its main goals are: (i) to investigate the effectiveness of combining textual and visual features for cross-lingual image retrieval; (ii) to collect and provide resources (e.g., data sets, topics, and relevance assessments) for benchmarking image retrieval systems; and (iii) to promote the exchange of ideas to help improve the performance of future image retrieval systems [60]. ImageCLEF consists of a number of tasks organized in different domains (e.g., medical image retrieval, historical archives, news photographic collections, and Wikipedia pages) and categories: ad hoc retrieval, object and concept recognition, and interactive image retrieval [60].

Two other relevant challenges in the context of VIR are: (i) The MediaEval Benchmarking Initiative for Multimedia Evaluation[10] and (ii) the ImageNet Large Scale Visual Recognition Challenge 2012 (ILSVRC2012).[11]

3.6   FEATURE EXTRACTION USING LIRE

LIRE implements many of the features introduced in this chapter [49, 50]. The most prominent global features available in LIRE are shown in Table 3.2. The characteristics listed for each feature are indicated by "yes" or "no" in the corresponding column, according to the following notation.

- Color (c): features that incorporate a straightforward "crisp" (not fuzzy) color information.

- Color distribution (cd): features that investigate how color pixels are related to each other (e.g., large blobs of the same color, or noise).

- Fuzzy color (fc): features that incorporate a fuzzy color scheme.

- Texture (t): features that investigate edges, gradients or other pure texture characteristics.

- Joint histograms (jh): features that combine different aspects of pixels, e.g., texture and color.

Revisiting the robustness discussion introduced earlier in this chapter, we can easily see that most of the color features (auto color correlogram, scalable color, simple color histogram) are robust to lossless rotation, and so are the Gabor and Tamura texture features. The texture features involving edge direction usually cannot tolerate rotations greater than $\pm 15°$. Features based on fuzzy color schemes (CEDD, FCTH, and JCD) are robust to scaling and rotation. Color layout and edge histogram are also robust to scaling. Robustness against cropping and translation is rarely achieved with global features. Hence, local features (Section 3.3) should be used if this is necessary.

[9] http://imageclef.org/
[10] http://multimediaeval.org/
[11] http://www.image-net.org/challenges/LSVRC/2012/index

Table 3.2: LIRE's global features and their main characteristics

Feature / Descriptor	c	cd	fc	t	jh
Auto color correlogram	yes	yes	no	no	no
CEDD	no	no	yes	yes	yes
Color layout	yes	yes	no	no	no
Edge histogram	no	no	no	yes	no
FCTH	no	no	yes	yes	yes
Gabor	no	no	no	yes	no
JCD	no	no	yes	yes	yes
Scalable color	yes	no	no	no	no
Simple color histogram	yes	no	no	no	no
Tamura	no	no	no	yes	no

In addition to the global features listed in Table 3.2, LIRE provides access to a SIFT implementation (from ImageJ[12]) and a SURF implementation (from jopensurf.[13]) Both local and global features implement the `LireFeature` interface described in Section 5.1. Global features can be extracted from `BufferedImage` instances with the `LireFeature.extract(BufferedImage)` method. The feature vector, a `double[]` array, can be accessed after extraction with the `LireFeature.getDoubleHistogram()` method (see also listing 5.1). The code for extracting CEDD from an image is shown in Listing 3.6. Extracting other features works the same way; all you have to do is to change the constructor in line 1 to the respective feature.

Listing 3.6: Example code for extracting the CEDD feature with LIRE

```
1  LireFeature feature = new CEDD();
2  BufferedImage image = ImageIO.read(new FileInputStream("img.jpg"));
3  feature.extract(image); double[] histogram = feature.getDoubleHistogram();
```

Local features cannot be extracted the same way as described earlier for global features, because there are many local features per image. A common approach within LIRE is to use the respective `DocumentBuilder` instances, `SurfDocumentBuilder` and `SiftDocumentBuilder`. They basically wrap the functionality for extraction provided by the underlying libraries. More information on how to deal with local features for content-based image retrieval can be found in Section 4.4.

SUMMARY

In this chapter we presented the basic building blocks of a content-based image retrieval system. We introduced commonly used color and texture features as well as state-of-the-art joint and hybrid

[12]http://rsbweb.nih.gov/ij/
[13]http://code.google.com/p/jopensurf/

features combining both color and texture. We also introduced two of the most popular local descriptors in use today (SIFT and SURF) and presented some of the most useful metrics to quantify the (dis)similarity between two images. Moreover, we offered a practical overview of the process of benchmarking VIR solutions, with many pointers to publicly available datasets and challenges posed to the multimedia research community.

PROBLEMS

3.1 Write Java code to implement a simple color histogram feature, with a number of bins of your choice. Normalize and quantize the bins.

3.2 Index the Ferrari test dataset with your custom-built feature from the previous problem and perform a few search operations with a distance function of your choice.

3.3 Extend your feature by implementing a joint histogram that combines your color bins with pixel rank. The neighborhood for computing the rank should be determined by a maximum L_1 distance of 2 to pixel $p = (x_p, y_p)$, so each pixel $q = (x_q, y_q)$, for which $L_1(p, q) = |x_p - x_q| + |y_p - y_q| \leq 2$, is taken into account.

3.4 Download the SIMPLIcity data set and evaluate your custom feature by determining $p_{at}(10)$ and the error rate.

CHAPTER 4

Indexing Visual Features

When users search a multimedia information system, they expect immediate response and relevant results. While the concept of *relevance*—in the case of CBIR, expressed in terms of the amount of visual similarity between the results and the query image—has been explained in previous chapters, the actual process of searching (and how to get results immediately) has not yet been discussed. In CBIR, a typical searching scenario uses a data structure called *index*. The basic process of searching an image repository and the role of the index are shown in Figure 4.1. Indexing an image consists of extracting low-level features from the image data and storing a representation of the extracted features into the index. Images are usually indexed "offline," i.e., before search time. The CBIR search process using a query-by-example (QBE) paradigm consists of presenting a query image to the system, indexing the example image ("online"), and performing a similarity search against the previously stored index. The results of the search process are usually presented as a ranked list of images, sorted in decreasing order of similarity to the query image.

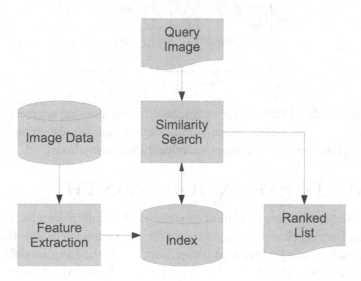

Figure 4.1: A typical CBIR search process operates on an index, which has been created before search time and provides a fast and efficient data structure for searching.

In text retrieval the use of an *inverted index* (or inverted list) is common practice. Essentially, an inverted index is an index data structure that stores a mapping from words to their locations in a set of documents. An inverted index is organized by a *term dictionary*, containing all the terms of all indexed documents. For each term there is a list indicating in which documents the respective term appears; these lists are called *posting lists*. At search time, only the terms used in the query need to be looked up in the term dictionary. The lists of documents referenced by the query terms are then used to compute the rank of each document occurring in at least one of the list elements. The relevance function is typically based on tf*idf and cosine coefficient, which conveniently work well together with the index structure of an inverted list.

Example 4.1

Table 4.1 gives an example of an inverted index. The first column is the term dictionary. The posting lists in the second column contain a pointer to each document in which the term appears and the number of occurrences of the term in that document, i.e., the term frequency. For example, the term *ferrari* appears twice in document d_1, once in document d_3, twice in document d_4, etc. Since it does not appear in document d_2, the ranked list of results will not include d_2.

Table 4.1: Example of an inverted index	
Term	**Posting List**
ferrari	$(d_1, 2), (d_3, 1), (d_4, 2), \ldots$
car	$(d_2, 1), (d_3, 2), \ldots$
fast	$(d_1, 1), (d_4, 1), \ldots$
maserati	$(d_3, 3), (d_4, 1), \ldots$
...	...

In the remainder of this chapter we discuss different indexing approaches of increasing degree of complexity and the associated computational implications. We also demonstrate how indexing is performed in LIRE (with support from underlying Lucene methods).

4.1   INDEXING: THE NAÏVE APPROACH

A simple approach to indexing consists of extracting and encoding the relevant image features before search time, and storing them in a database, in a file, or in memory, to have them readily available at search time. This basically leads to a list (or set, since there is no intrinsic order) of data entities, that includes: (i) a pointer to the image (file name, URI, etc.) and (ii) the extracted low-level features (encoded in a suitable way). At search time this list (or set) is traversed linearly and similar images are collected for a result set (which must subsequently be ordered according to their relevance relative to the query). During the traversal process, each and every data entity has to be: (i) read from disk;

(ii) decoded; (iii) compared to the query image's equivalent encoded features; and (iv) eventually added to the result set. For n data entities this results in a computational complexity of $O(n)$.

It is important to note that computational complexity—especially for small values of n—depends strongly on constants that are not apparent in the $O(n)$ formulation. With linear search, common hidden constants include the following.

- **Properties of the actual medium in which the data entities are stored.** SSDs, HDDs, main memory, and network storage all have very different characteristics influencing the search process.

- **Decoding of data entities.** The data entities in the resulting list (or set) are often encoded (i.e., compressed) features. There is an inherent tradeoff involved in the process: while decoding takes time, compression, on the other hand, can allow for faster transfers between different media, e.g., from HDDs to main memory.

- **Object access and manipulation overhead.** In high-level programming languages, such as Java and C#, each and every object has its properties stored in a header (16 bytes for a `java.lang.Object` instance on a 64-bit Oracle VM). This header eventually adds to the size of the data entities, especially if you use nested objects and arrays of objects.[1]

4.1.1   BASIC INDEXING AND LINEAR SEARCH IN LIRE

LIRE serves as good example for linear search in Java. In LIRE, linear search is the default approach, mainly to reduce complexity of usage for novice users, but also due to its satisfactory performance for small and moderate-size image repositories. Based on Lucene, the traversal of the data entities is rather fast and typically leads to a search time $t_s < 0.5$ seconds for 100,000 images.[2] For indexing, LIRE provides several implementations of the interface `DocumentBuilder`, that takes an instance of the `BufferedImage` object along with a file name and creates a Lucene `Document`. It's up to the user of the API to create an instance of the Lucene `IndexWriter` and write the document to disk, to the network, or store it in main memory.

Listing 4.1 shows the main traversal of Lucene `Document` instances in the abstract class `GenericFastImageSearcher`, which corresponds to the main implementation for linear search in LIRE. Basically, each and every document in the index is read after checking if the document has been deleted or not (cp. lines 1 and 4). In Lucene, deleted documents reside in the index unless the index is re-optimized; only then deleted entries are removed from the index. If a document is there, it is read from the index (cp. line 5) and its distance to the query feature is computed (cp. line 6) using the `getDistance(...)` method, which is explained in detail in Listing 4.2. The intermediate result list `this.docs` is limited to a number of candidate documents `maxhits` (cp. line 11). Based on the maximum distance of the result set and the maximum number of candidate results it is decided if the

---

[1]Note that primitive types in Java have smaller headers.
[2]Recent tests on a quad-core AMD Windows 7 PC with 8 GB RAM and a 256 GB SSD have shown a search time of 0.268 seconds for 100,000 images using the CEDD feature.

current document should be included in the intermediate result list this.docs. If this is the case the document gets added and eventually the last document is removed to not exceed the maximum size (cp. lines 11–18).

Listing 4.1: Part of the main search method in LIRE as implemented in the abstract class GenericFastImageSearcher

```
1    Bits liveDocs = MultiFields.getLiveDocs(reader);
2    for (int i = 0; i < docs; i++) {
3        // check if the document isn't flagged as deleted
4        if (hasDeletions && !liveDocs.get(i)) continue;
5        Document d = reader.document(i); // Test Comment
6        float distance = getDistance(d, lireFeature);
7        assert (distance >= 0);
8        if (overallMaxDistance < distance)
9            overallMaxDistance = distance;
10       if (maxDistance < 0) maxDistance = distance;
11       if (this.docs.size() < maxHits) {
12           this.docs.add(new SimpleResult(distance, d));
13           if (distance > maxDistance)
14               maxDistance = distance;
15       } else if (distance < maxDistance) {
16           this.docs.remove(this.docs.last());
17           this.docs.add(new SimpleResult(distance, d));
18           maxDistance = this.docs.last().getDistance();
19       }
20   }
```

Listing 4.2 shows the getDistance(Document, LireFeature) method, which is critical to the speed of the linear traversal. Note that there are no local variables created within the method, since in Java creation of a variable takes time, leads to allocation of memory, and ultimately triggers garbage collection, which is known to block processes and to reduce speed significantly. The variable LireFeature cachedInstance has been initialized at instantiation time of the GenericFastImageSearcher (cp. line 4). All features stored in Lucene documents are byte encoded, as this is the fastest way to store to and read from an index. Implementations of the interface LireFeature have to implement the method setByteArrayRepresentation(byte[], int, int) (cp. lines 4-7) in an efficient way to allow for fast search.

Listing 4.2: getDistance(...) search method in LIRE as implemented in the abstract class GenericFastImageSearcher

```
1    protected float getDistance(Document document, LireFeature lireFeature) {
2        if (document.getField(fieldName).binaryValue() != null
3                && document.getField(fieldName).binaryValue().length > 0) {
4            cachedInstance.setByteArrayRepresentation(
5                document.getField(fieldName).binaryValue().bytes,
6                document.getField(fieldName).binaryValue().offset,
7                document.getField(fieldName).binaryValue().length);
8            return lireFeature.getDistance(cachedInstance);
9        } else {
10           logger.warning("No feature stored in this document! "+descriptorClass.getName());
```

```
11      }
12      return 0f;
13  }
```

At this point the question "Why does LIRE use the text retrieval engine Lucene for linear search?" can be easily answered. First of all, Lucene does not employ a client-server architecture by default. Developers can just use the library and give it a directory for an index to start with. Still, it is scalable enough that an index can be put into a database, into main memory, or distributed over a network. Second, compared to a database, Lucene has a very small overhead. There is no user management, transaction management, or advanced features typically found in big database management systems. Third, Lucene takes care of concurrency at search and index time. The index is locked while writing and for searching the number of IndexSearcher and IndexReader instances is controlled by Lucene. The only inconvenience for LIRE is that we have to check for deleted documents manually as they are first flagged and removed only later; all searches in between flagging and actual removal may inadvertently turn them up. Finally, Lucene virtualizes index access and allows for easy switching between different implementations including memory mapped indexes, main memory indexes, and file system indexes. For the latter, even implementations optimized for different operating systems and Java VMs are included and selected automatically.

4.2   NEAREST-NEIGHBOR SEARCH

While the naïve approach works fine up to a certain number of images, scalability in terms of data set size is still unsatisfactory. Over the years, different approaches for search with *better than linear* runtime complexity have been developed, tested and employed in application scenarios. Typically, these approaches provide a trade-off between runtime complexity and accuracy of results, which means that the results are approximate. There are many such approaches to choose from, which poses a critical problem for developers, namely that of selecting the right approach for a given search problem. Notable from a historical point of view are the dimensionality reduction approaches, with *FastMap* [24] being one of the most prominent. These approaches are based on the idea that spatial indexes, such as the R-tree [29] or the R*-Tree [8], provide efficient access to the nearest neighbors of a point in multidimensional space.

Figure 4.2 shows an example in two-dimensional space. Diamonds denote documents in the repository, whereas the query is indicated by the star. A spatial index would return all documents within a given radius as shown by the dashed circle. Note that the radius for nearest neighbor search can be given as a parameter at search time: the larger the radius, the more neighbors will be taken into account.

The biggest problem with spatial indexing is the *curse of dimensionality*. Complexity virtually explodes with the number of dimensions and it is well known that spatial indexes only work well with less than ten dimensions [9]. Since extracted low-level image features are usually of high-dimensionality (in the tens or hundreds), the number of dimensions has to be reduced in order to cope with the problem.

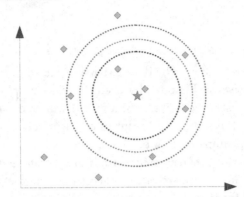

Figure 4.2: Spatial search example in 2D: documents are represented as diamonds, the query is indicated by a star, and the radius of the dashed circle represents how many nearest neighbors will be taken into account during the search process (in this case, 2, 3, and 6 neighbors, respectively, as we move from the smallest to the largest circle).

Dimensionality reduction always comes with a loss of information. The most notable requirement for a dimensionality reduction algorithm is that the original pair-wise distances between images need to be retained, i.e., if two images are distant from each other, they still should be distant in relative numbers after the reduction step. The same goes for pairs of images with small distances. Therefore, it is a minimization problem with the objective function $z = \sum_{i,j=0}^{n} |d(I_i, I_j) - \bar{d}(\bar{I}_i, \bar{I}_j)|$ minimizing the error between the original distance function $d(I_i, I_j)$ of two images I_i and I_j and the distance function after dimensionality reduction $\bar{d}(\bar{I}_i, \bar{I}_j)$ between the images $\bar{I}_i, \bar{I}_j$ projected to a lower dimensional space.

Directly minimizing such an error function iteratively can be done with a *force directed placement* or *spring embedding* algorithm. Both define forces between data points along with a minimum attraction and retraction. Algorithm 1 gives a general approach. Note that computation of the movement vector includes comparison to all other data points. The consideration of a minimum attraction ensures that data points won't drift away too far from each other and a minimum retraction ensures that data points keep a minimum distance, so that no two points end up at the same coordinates. Force directed placement has a computational complexity of $O(n^3)$. Therefore, it is—in its original form—not feasible for the use of dimensionality reduction in image indexing for large image sets.

Faloutsos and Lin [24] proposed an approach to multidimensional scaling called *Fastmap*. Fastmap's actual complexity is $O(nk)$, whereas n is the number of data points and k is the number of dimensions. Basically, for each dimension, Fastmap takes two pivot elements and interpolates all other data points in between. A general description is given in Algorithm 2.

Algorithm 1 Force directed placement algorithm.

Input : Set of data points along with their pair-wise distance.
Output : Coordinates of data points in lower dimensional space.

1: Place all data points in target space randomly.
2: **repeat**
3:      **for all** data points **do**
4:          Compute movement vector based on original distances to all other data points.
5:      **end for**
6:      **for all** data points **do**
7:          Move data point along movement vector.
8:      **end for**
9:      Compute error function z.
10: **until** $z <$ a given threshold

Algorithm 2 Fastmap.

Input : Set of data points along with their pair-wise distance.
Output : Coordinates of data points in lower dimensional space.

1: **for all** dimension k **do**
2:      Choose two pivot elements p_1^k, p_2^k from the data points.
3:      **for all** data points d **do**
4:          Place d in k by interpolation based on the distance of d to p_1^k, p_2^k
5:      **end for**
6: **end for**

While spatial indexing and dimensionality reduction were important milestones for visual information retrieval, there was still need for further improvement. Problems such as loss of information during the dimension reduction step as well as management of spatial indexes remained and must be addressed. Additionally, trade-offs between search or indexing time and precision or recall turned out to be challenging as spatial indexes tend to become significantly slow with increasing number of dimensions. Therefore, different alternative methods have been developed over the years. Three of the most prominent methods are described in the remainder of this chapter: hashing, metric space indexing, and bag of visual words.

4.3 HASHING

Hashing describes the process of reducing a large amount of data (e.g., a large text document or a video) to a single number, called *hash*. Hashing functions used in security applications, such as SHA-2, MD5, etc. [3] [66], are by definition: (i) collision-free, which means that no two documents different in at least one bit should result in the same hash; and (ii) extremely hard to invert, so that documents cannot be re-constructed from their hashes, i.e., a hashing function behaves as a trapdoor function. In the context of information retrieval, however, these two aspects of hashing are handled differently. Inversion is typically not a problem to which much thought is given. It basically does not matter if the original content of the document can be inferred from its hash. Collisions, however, are carefully constructed as follows: collision between two documents hashes $h(d_i) = h(d_j)$ should occur if, and only if, the documents d_i, d_j are similar, i.e., the similarity $s(d_i, d_j) < t$ where t is a suitable threshold.

4.3.1 LOCALITY SENSITIVE HASHING

Locality Sensitive Hashing (LSH) [17] is a hashing technique that uses random projections of the feature vectors. Multiple random projections are combined into a hash value. The basic idea behind LSH is to map a feature vector $v \in \mathbb{R}^d$ to an integer h_i as follows:

$$h_i(v) = \left\lfloor \frac{a_i v + b_i}{w} \right\rfloor , \qquad (4.1)$$

where $a_i \in \mathbb{R}^d$ is a vector of random independent components from a normal distribution, $w \in \mathbb{R}^+$ is a constant influencing the granularity of the results, and $b_i \in [0, w)$ is a random, uniformly distributed number. A k-tuple of these integers h_i with $i \in \{1, 2, 3, ..., k\}$ defines a composite hash-value

$$h(v) = (h_1, h_2, h_3, ..., h_k).$$

Note that a_i, b_i, and w, as well as k, have to be chosen once and are not meant to change for an index. For search we assume that if $m < k$ hashes of the hash tuples match two images, then the images are candidates for nearest neighbors. Note that this specific version has been designed for L_2 distance. There are different types of LSH hashing functions to serve different needs. A more detailed description and further pointers are given in [70].

For implementation we need to create a hash bundle from randomly generated numbers for use in indexing and search. For a_i, normally distributed random numbers are needed. A common method to draw normally distributed random numbers is the Box-Muller [10] transform (see Eq. 4.2) where $k_1, k_2 \in (0, 1]$ are uniformly distributed random numbers. The resulting z_0, z_1 are independent random variables with standard normal distribution:

$$\begin{aligned} z_0 &= \sqrt{-2 \ln k_1} \cos(2\pi k_2) \\ z_1 &= \sqrt{-2 \ln k_1} \sin(2\pi k_2) . \end{aligned} \qquad (4.2)$$

Algorithm 3 gives a general scheme for creating the hash bundles. Hash values for a low level feature are computed based on the histogram vector using Eq. 4.1. Note that the same hash bundle that was created at index time is needed at search time too.

Algorithm 3 Creating a hash bundle for approximating L_2 distance.

Input : Number of dimensions, number of hash functions in the bundle, w.
Output : Hash bundle.

1: **for all** hash functions h_i in h **do**
2:     Draw $b_i \in [0, w)$ (uniform distribution).
3:     **for all** dimensions d **do**
4:         Draw $a_{i,d}$ (normal distribution, e.g., Box-Muller transform).
5:     **end for**
6: **end for**

Listing 4.3 gives a Java implementation of Algorithm 3 based on the LIRE code. First, the number of dimensions and the size of the hash bundle are serialized (cp. lines 3-4). Then, b_i is drawn for all hash functions (cp. lines 5-6). Finally, all a_i are drawn. The method `drawNumber()` returns a variable from a standard normal distribution based on the Box-Muller transform (see Eq. 4.2). To read from the generated file one has to employ the respective methods of `ObjectInputStream` in the same order.

Listing 4.3: Creating a hash bundle and serializing it

```
1   ObjectOutputStream oos = new ObjectOutputStream(new
2        FileOutputStream("hashes.obj"));
3   oos.writeInt(dimensions);
4   oos.writeInt(numFunctions);
5   for (int c = 0; c < numFunctions; c++)
6       oos.writeDouble(Math.random()* w);
7   for (int c = 0; c < numFunctions; c++)
8       for (int j = 0; j < dimensions; j++)
9           oos.writeDouble(drawNumber());
10  oos.close();
```

4.3.2 METRIC SPACES APPROXIMATE INDEXING

While hashing has proven its value over the years, it still has a drawback: the decision on which images are nearest neighbor candidates is binary, i.e., typically no ranking of the candidate images is possible. Therefore, as soon as the parameters are chosen—which happens at indexing time—there is no way to retrieve additional candidates for nearest neighbors. Moreover, the main data structure to be employed is a hash table, which is not easy to handle efficiently when the size of the table exceeds the available main memory. Since similar problems occur in text retrieval, advancements in that field have shown multiple ways to handle large and distributed inverted lists and many practical implementations of inverted lists, such as Lucene, are available.

An approach suitable for fast approximate indexing and search in metric spaces based on inverted lists has been presented in [5]. A running prototype with an index featuring more than 100 million images can be found on G. Amato's homepage.[3] The basic idea is that objects in a metric space, which are close to each other, *see* other objects in a similar way. More specifically, objects that are close to one another are more likely to have a similar list of nearest neighbors. To give a high-level example, consider two people p_1, p_2, living near the Eiffel tower and the Louvre and a third person p_3 living near the Colosseum and St Peter's Basilica. Most likely, p_1 and p_2 live near to each other in relation to their distance to p_3, who lives far away from p_1 and p_2.

To employ this idea in a retrieval system, we first need to introduce a metric distance function $d(I_i, I_j)$ capable of computing the distance between two images I_i, I_j, based on neighboring objects. Furthermore, a set of reference objects $RO = \{r_1, r_2, r_3, ..., r_k\}$, just like *fixed stars*, is needed to provide a dedicated set of neighbors. Now, for each image I_i an ordered list containing all reference objects RO is created. The list order is determined by the distance of I_i to the elements of RO starting with the nearest reference object.

This leads to a *perspective-based space transformation* where each image I_i is described by an ordered list of reference objects. Distance in this new space is determined by the difference between those lists based on *Spearman's Footrule Distance* [21]. Let $p_i(r)$ with $r \in RO$ be the position of the reference object r in the ordered list of reference objects describing Image I_i. Spearman's Footrule Distance d_s then is

$$d_s(I_i, I_j) = \sum_{r \in RO} |p_i(r) - p_j(r)| \,. \tag{4.3}$$

Example 4.2   Figure 4.3 gives a simple example resulting in the following ordered lists of reference objects, assuming that the distance function d is L_2, and the following calculation for Spearman's Footrule Distance:

$$I_1 \rightarrow (r_4, r_1, r_3, r_2)$$
$$I_2 \rightarrow (r_1, r_2, r_3, r_4)$$
$$I_3 \rightarrow (r_3, r_2, r_1, r_4)$$
$$d_s(I_1, I_2) = |1 - 4| + |2 - 1| + |3 - 3| + |4 - 2| = 6$$
$$d_s(I_1, I_3) = |1 - 4| + |2 - 3| + |3 - 1| + |4 - 2| = 8$$
$$d_s(I_2, I_3) = |1 - 3| + |2 - 2| + |3 - 1| + |4 - 4| = 4 \,.$$

As can be easily seen, the relative distances are preserved, i.e., I_2 is closest to I_3, and I_1 is farthest from I_3. The distance function also reflects that I_1 is closer to I_2 than to I_3, which also appears in the diagram.

---

[3]http://www.nmis.isti.cnr.it/amato/

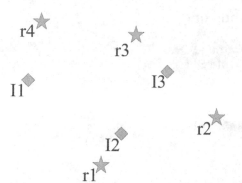

Figure 4.3: Example for a perspective based space transformation with three objects (I_1, I_2, and I_3) and four reference objects (r_1,..., r_4). See text for details.

For implementation purposes, this operation can be done in an inverted list. The inversion of the data for each r_k leads to a posting list of images I_i along with the position $p_i(r_k)$ for each image I_i:

$$
\begin{aligned}
r_1 &\rightarrow (I_1, p_1(r_1)), (I_2, p_2(r_1)), \ldots \\
r_2 &\rightarrow (I_1, p_1(r_2)), (I_2, p_2(r_2)), \ldots \\
&\ldots \\
r_k &\rightarrow (I_1, p_1(r_k)), (I_2, p_2(r_k)), \ldots .
\end{aligned}
\tag{4.4}
$$

The most critical point is that one needs to reduce: (i) the size k of RO, i.e., the overall number of reference objects; and (ii) the length of the list identifying images I_i in the transformed space, i.e., the number of reference objects representing a single image. Therefore, the posting lists are reduced to list just the L-nearest reference objects, instead of all k of them. This leads to a much sparser posting list than the one shown in Eq. 4.4.

At search time the L-nearest reference objects of the query image I_q are computed and the list is used to: (i) access the posting lists of each reference object; and (ii) compute Spearman's Footrule Distance for each image found in at least one posting list. Search run time complexity is, therefore, dependent on: (i) the number of images in the posting list (k); and (ii) the number of reference objects describing a single image (L). Based on their experience and the test results from their evaluation, the authors of [5] proposed using a reference object set with $|RO| \geq 2\sqrt{N}$ with N being the total number of images. In the same publication it has been shown that reference objects chosen randomly led to robust and good results in search performance.

An implementation has to use two indexes: one for the reference objects (RO) and another one for the actual indexed objects, the perspective based index (*PB Index* in Figure 4.4. For indexing (see left portion of Figure 4.4), the reference objects are chosen randomly from the data set. In a second step, the reference object index gets searched for each image to be indexed and the list of the

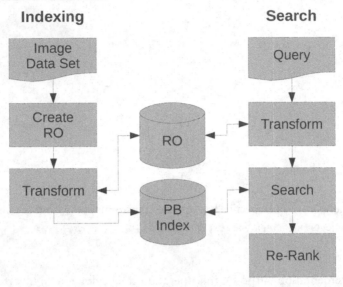

Figure 4.4: Flow diagram for (i) indexing a data set and (ii) searching the indexed data based on two indexes; the reference object index (RO) and the perspective based index (PB Index).

first k reference objects is stored in the perspective based index. At search time (see right portion of Figure 4.4) the feature extracted from the query image is used to query the reference object index. The ranked list of the k nearest reference objects is used again to query the perspective based index. Results are then re-ranked and presented to the user.

4.4   BAG OF VISUAL WORDS

A very popular method for indexing large repositories of images is the *bag of visual words* (BoVW) approach, originally proposed in [73]. The basic idea is to use metaphors, methods, and schemes that have been successful in the area of text retrieval for visual information retrieval. Local features, such as SURF, SIFT, and MSER, as already discussed in Section 3.3, provide means to identify and describe key points over different images. They allow for tracking of objects, identification of overlap in images, and object recognition in visual media. Based on the assumption that similar key points are to be found with visually similar image patches, we can state that key points can be used as terms to describe images in a way that is analogous with words (terms) used to describe documents. We can further state that images having the same key points will probably contain similar scenes, objects, etc.

However, while terms are typically occurring in multiple documents, i.e., the same words are used in different texts, identical SIFT, and SURF features are not to be found in more than one image, but similar key points have similar feature vectors, i.e., feature vectors with a minimum pair-

Image 1   Image 2   Image 3

Figure 4.5: Simplified example for indexing with visual words. All three images show a different composition of objects.

wise distance. To employ the idea of terms in a text index we need to combine several local features to a single *visual word*. A common approach is to take a significant part of available key points from a data set and cluster them. The resulting clusters are then considered *visual words*. Hence, all local features that fall into the same cluster represent the same visual word. Figure 4.5 shows a simplified example for three images and different compositions of objects. For the sake of simplicity, let's assume that after the clustering of local features each object (cloud, flower, sun, moon, tree) corresponds to exactly one visual word. One can then build an inverted list like this:

$$
\begin{aligned}
cloud &\to (I_1, 1), (I_3, 1) \\
flower &\to (I_1, 2), (I_3, 2) \\
moon &\to (I_2, 1), (I_3, 1) \\
sun &\to (I_1, 1) \\
tree &\to (I_2, 1), (I_3, 1) \ .
\end{aligned}
\tag{4.5}
$$

Figure 4.6: Simplified process diagram for indexing with bag of visual words.

With the use of an inverted list as data structure (cp. Eq. 4.5), efficient search can then be implemented. However, the actual implementation has several quirks that need to be addressed.

Figure 4.6, shows a simplified version of the indexing process using bag of visual words. For each of the steps there are several aspects to consider.

Local Feature Extraction. There are many local feature extraction methods currently available (cp. Section 3.3). Note that many classical methods, e.g., SIFT and SURF, operate on gray images and do not take color information into consideration. Moreover, the resolution of the images to be indexed influences the number of local features found: larger images allow for indexing in greater detail but lead to many more local features. Re-scaling images to a similar size at index time is, therefore, recommended. It should also be noted at this point that features of an eventual query image have to be extracted at search time, so the faster the local feature extraction method, the better.

Clustering Local Features. Essentially, any clustering algorithm capable of creating a sense of structure from the data can be employed for generating the visual words, i.e., building a *visual vocabulary*. A very popular choice is the k-means clustering algorithm [38], for several reasons: (i) it is easy to implement (also for parallel processing); (ii) many implementations are readily available; and (iii) the algorithm's characteristics are well known. However, experiments with fuzzy and hierarchical clustering, among other variations, have been done (see e.g. [40]). Note that for large repositories not all of the local features are clustered. A random sample of images is selected and only local features associated with the selected images are clustered. The actual number of local features to cluster for the visual vocabulary is an issue of current research. In the literature, vocabulary sizes of 100–10,000 visual words (clusters) are reported.

Indexing Images. Having created a visual vocabulary, indexing all images is an issue of mapping available local features to existing clusters. While this depends on the actually employed clustering algorithm we assume there is a cluster center, e.g., a mean or medoid, that represents the cluster. For each local feature the vocabulary is searched for the nearest cluster center and the corresponding visual word is added to the image the local feature belongs to. This process is known as *hard assignment*, where one local feature can only be assigned to a single visual word. *Soft assignment*, in contrast, allows for fuzzy assignment of a local feature to multiple visual words based on a membership function. It has been shown that soft assignment can compensate small visual vocabulary sizes [40]. Note that the visual vocabulary is also required at search time as local features are extracted from the query image and the visual words are assigned.

4.4.1 BAG OF VISUAL WORDS USING LIRE

LIRE provides bag of visual words (BOVW) indexing and search capabilities. Usage and adaptation of the LIRE library will be discussed in the next section. Basically, it is recommended to use whatever tools are available to the developer. For local features there are implementations of SIFT (e.g., JavaSIFT[4]) and SURF (e.g., jopensurf.[5]) Also, OpenIMAJ [34] provides multiple ways to extract

---

[4]http://fly.mpi-cbg.de/~saalfeld/Projects/javasift.html
[5]http://code.google.com/p/jopensurf/

local features including SIFT and MSER. The most convenient method is the use of jopensurf, since the keypoints and their descriptors can be extracted in just two lines (see Listing 4.4).

Listing 4.4: Extracting SURF interest points and their descriptors with jopensurf

```
1   BufferedImage image =
2           ImageIO.read(new FileInputStream("image.jpg"));
3   Surf s = new Surf(image);
4   List<SURFInterestPoint> interestPoints =
5           s.getFreeOrientedInterestPoints();
```

Clustering of local features can be done using the k-means clustering algorithm. Basically, there are two ways to do the clustering. One option is to use a tool such as Weka [30] or R[6] to do the clustering externally. Then the extracted key point descriptors must be prepared in a data file format supported by the respective tool. Another option consists of integrating the clustering functionality in the source code either with a library or with a simple implementation of k-means as shown in Algorithm 4.

Algorithm 4 Simple k-means clustering implementation

Input : Set of descriptors D and number of clusters k.
Output : Clusters and cluster means.

1: Initialize k empty clusters $C = \{c_1, ...c_k\}$.
2: **for all** Clusters $c \in C$ **do**
3:     Assign random mean $m \in D$ to c.
4: **end for**
5: **repeat**
6:     **for all** descriptors $d \in D$ **do**
7:         Assign d to c_j with minimum distance to the cluster mean m_j of c_j.
8:     **end for**
9:     **for all** Clusters $c \in C$ **do**
10:         Re-compute cluster mean of c.
11:     **end for**
12: **until** Re-assignment of descriptors reaches a minimum

Having created the visual vocabulary one can then create a *local feature histogram* encoding which visual word (cluster) occurs with what frequency in the image. A histogram such as $(3, 2, 0, 0, 0, 1)$ would, for instance, indicate that the visual word v_1 occurs three times in the indexed image, v_2 two times, and v_6 one time. Visual words $v_3 - v_5$ are not assigned to the example image. This local feature histogram can then be used to index the documents in an inverted list. In

[6]http://www.r-project.org/

LIRE, the histograms are encoded as text and put into Lucene. For the example above, the resulting text would be "*v1 v1 v1 v2 v2 v6*." Using a Lucene `IndexSearcher` and `QueryParser` the indexed documents can be retrieved with similarly encoded query strings. However, as it is well known that tf*idf does not work well on visual words, one should overwrite the default Lucene similarity function as shown in Listing 4.5. This example just uses a logarithmic weight on the raw term frequency.

Listing 4.5: Sample similarity function for Lucene replacing the default tf*idf similarity implementation for bag of visual words

```
1   private static class MySimilarity extends DefaultSimilarity {
2       public float tf(float freq) {
3           return (float) Math.log(freq);
4       }
5
6       public float idf(int docfreq, int numdocs) {
7           return 1f;
8       }
9
10      public float queryNorm(float sumOfSquaredWeights) {
11          return 1;
12      }
13
14      public float computeNorm(String field, FieldInvertState state) {
15          return 1;
16      }
17  }
```

SUMMARY

In this chapter, we discussed common indexing strategies for image search and we have given directions to implement them in Java. The most intuitive method—linear search—involves traversal of the whole data set and is therefore the slowest of the proposed approaches. However, with thoughtful implementation, a reasonable performance on small to medium sized data set, i.e., up to 100,000 images, can be achieved. Faster indexing strategies typically involve a trade-off between search time and search performance, expressed in terms of precision and recall. They all return just a set of candidates, that needs to be ranked to provide an ordered list of results.

Each approach has benefits and drawbacks. Hashing approaches work well for near duplicate search and features that work with L_2 distance (or the respective distance function the hash family is constructed for), but do not perform well if search needs to go beyond identifying images that are only similar to a certain degree. Indexing in metric spaces works well with global features but needs a specialized inverted index implementation featuring Spearman's Footrule Distance. The bag of visual words approach is common practice, but has a complicated tool chain including clustering, random sampling and multiple parameters, e.g., which local features, clustering, and assignment strategy to use, the size of the visual vocabulary, etc. Therefore, finding the right configuration may prove difficult and application-dependent.

In conclusion, there is no "one-size-fits-all" solution. For smaller projects, linear search produces least overhead. For larger projects, developers are advised to consider search requirements, benefits and drawbacks of the different methods. Basic decisions include: determining whether local or global features are needed and whether search needs to find just near duplicates or also similar images, e.g., images with similar colors but different textures or images showing similar buildings, etc.

PROBLEMS

4.1 Implement LSH for the Ferrari data set. Use Listing 4.3 to create the hash bundle and apply the hash function to a feature of your choice extracted from the Ferrari data set. The result should be a hash table, where all the hashes are stored for all images.

4.2 Use the hash table generated in your solution to the previous exercise to test the *precision at three* of your LSH application to linear search with the same feature.

4.3 Implement indexing and search based on SURF and bag of visual words and test the *precision at three* on the Ferrari data set. Assume that the directories indicate categories (red, black, yellow, white) and each image from a category serving as a query image should return all others of the category.

CHAPTER 5

LIRE: An Extensible Java CBIR Library

LIRE is a Java library designed to enable quick deployment of VIR solutions. It can be integrated into existing projects or used to build image retrieval applications from scratch. LIRE is based on Lucene, a text search engine which provides capabilities such as: inverted indexing, search, and fast random access to text indexes.

From a developer's perspective, LIRE tries to hide the complexity of visual information retrieval as much as possible. There are very few parameters to be set and even fewer choices to be made. `DocumentBuilder` classes provide easy access to different low-level features and wrap the use of Lucene, which is used for indexing. `ImageSearcher` classes allow for search and retrieval based on single query images or already indexed documents. *Extensibility* is a main feature of LIRE. By implementing a simple interface—essentially by taking care of serialization of features within the Lucene index—developers can provide their own low-level features and distance metrics and share them as open source code with the community.

LIRE has seen its first release in February 2006. Since then it has been downloaded more than 26,000 times (as of December 2012). This does not include downloads from SVN or from the LIRE website[1] or blog.[2] LIRE has been continuously extended and maintained through the years. In 2008, students from a course on Multimedia Information Systems at Klagenfurt University contributed Java code for Tamura and Gabor features. In 2009, Savvas Chatzchristofis added the joint histogram descriptors CEDD and FCTH, which are currently the global features of choice for most practical applications. In 2010, a major re-write of the serialization of feature descriptors resulted in a noticeable performance increase and in 2011 another re-write of core functions simplified the use of local descriptors and index management of bag of visual words indexes. The last major change was the adaptation to Lucene 4.0 in 2012.

Besides core functions and features, several smaller—but nevertheless interesting—methods have been implemented in LIRE. Examples include: an estimator for visual attention in images, a parallelized version of the k-means clustering algorithm, a parallel indexer for large file repositories, a benchmarking suite based on the SIMPLIcity data set, and latent semantic indexing. Additionally, a desktop Java demo tool called LireDemo (see Section 1.2) has been developed, which allows for easy and interactive tests and experiments to be performed on arbitrary image collections.

[1]http://www.lire-project.net/
[2]http://www.semanticmetadata.net/lire/

Figure 5.1: LIRE building blocks. An image corresponds to exactly one Document, which contains between 1 and N low-level features from the image.

5.1   ARCHITECTURE AND LOW-LEVEL FEATURES

The main tasks of LIRE are: (i) to extract low-level features from images; (ii) to index the low-level features; and (iii) to perform search in a database of indexed features.

The main building blocks of LIRE are the following.

- **Images.** LIRE makes use of the `java.awt.image.BufferedImage` class to read pixel data. Image files are read with `ImageIO.read(InputStream)`. Therefore, LIRE only supports image files supported by the Java VM libraries. To support reading of bitmap images stored in other formats, we recommend using ImageJ[3] or Sanselan[4] (see Listing 3.2 for an example).

- **Lucene.** LIRE uses the text search engine Lucene to maintain the index. Lucene allows for indexing of text and metadata in combination with the low-level features, has a small footprint, is easy to use and manage, and has proven to be fast enough for many different scenarios. An image file with all its features is typically reflected by a single Lucene `Document`.

- **Low-level features.** LIRE extracts features from images and assigns the resulting feature vectors to the corresponding `Document`.

The architecture of LIRE can be understood with the help of a simple entity-relationship diagram (Figure 5.1). An image corresponds to exactly one Lucene `Document` instance, which contains between 1 and N low-level features from the image. Low-level features are encapsulated in classes implementing the `LireFeature` interface as given in Listing 5.1. Each low-level feature implementation must provide: (i) its own feature extraction algorithm; (ii) its own serialization code (with support for byte arrays and Java Strings) ; and (iii) a method to compute a distance function suitable for the specific feature. An additional method, which returns an array of `double`s, ensures that the feature works with algorithms that are implemented independently from the actual feature, e.g., the latent semantic analysis (LSA) implementation `LSAFilter`.

---

[3]http://rsbweb.nih.gov/ij/
[4]http://commons.apache.org/imaging/

Listing 5.1: LireFeature interface

```
1   public interface LireFeature {
2       public void extract(BufferedImage bimg);
3       float getDistance(LireFeature feature);
4       public double[] getDoubleHistogram();
5       // byte[] serialization
6       public byte[] getByteArrayRepresentation();
7       public void setByteArrayRepresentation(byte[] in);
8       public void setByteArrayRepresentation(byte[] in, int offset, int length);
9       // String serialization
10      java.lang.String getStringRepresentation();
11      void setStringRepresentation(java.lang.String s);
12  }
```

Features can be added easily by implementing the `LireFeature` interface (cp. Listing 5.1). Listing 5.2 shows a sample implementation of the interface. The sample feature in this case has a two-bin histogram, containing width and height of the image.[5] Each feature needs an empty constructor, because features are created inside LIRE with `Class.newInstance()` (cp. lines 5–9). The main logic for extracting the features from the image can be found in the extraction method (cp. lines 11–14). Serialization of the feature has to be implemented to support its use within LIRE and Lucene. There are serialization utility classes available that cover the standard cases (cp. lines 16–26). The distance measure is specific to each implementation. One must take care of casting and eventual errors in typecasting, which is not done in this example, in one's code. Common distance functions are implemented in the `MetricUtils` class (cp. lines 32–35). String serialization is part of the interface due to historical reasons. However, since 2011 integration in Lucene indexes is based on `byte[]` arrays and therefore, this type of serialization is not supported in newer low-level features—and our example reflects that by throwing an exception (cp lines 37–44).

Listing 5.2: Sample LireFeature interface implementation

```
1   public class SampleFeatureImplementation
2           implements LireFeature {
3       double[] hist;
4
5       public SampleFeatureImplementation() {
6           this.hist = new double[2];
7           hist[0] = -1;
8           hist[1] = -1;
9       }
10
11      public void extract(BufferedImage bimg) {
12          hist[0] = bimg.getWidth();
13          hist[1] = bimg.getHeight();
14      }
15
16      public byte[] getByteArrayRepresentation() {
17          return SerializationUtils.toByteArray(hist);
18      }
19
```

---

[5]Note that this is not necessarily a useful feature in terms of retrieval, but still a convenient example for implementation.

```
20      public void setByteArrayRepresentation(byte[] in) {
21          hist = SerializationUtils.toDoubleArray(in);
22      }
23
24      public void setByteArrayRepresentation(byte[] in, int offset, int length) {
25
26          hist = SerializationUtils.toDoubleArray(in, offset, length);
27      }
28
29      public double[] getDoubleHistogram() {
30          return hist;
31      }
32
33      public float getDistance(LireFeature feature) {
34          return (float) MetricsUtils.distL2(hist,
35              ((SampleFeatureImplementation) feature).hist);
36      }
37
38      public String getStringRepresentation() {
39          throw new UnsupportedOperationException(
40              "Not implemented due to performance issues.");
41      }
42
43      public void setStringRepresentation(String s) {
44          throw new UnsupportedOperationException(
45              "Not implemented due to performance issues.");
46      }
47  }
```

Note that feature implementations are critical in terms of performance. The indexing process often involves extracting features from tens or even hundreds of thousands of images. With this number of images it makes a huge difference to extract a feature in 25 or 250 ms. Moreover, feature extraction at search time can be a critical task. If search is based on an example image one has to add the extraction time to the overall search time. Additionally, the distance function implementation is executed for each image in the index at search time in the case of a linear search. So one has to be careful to avoid code that influences performance significantly. Common pitfalls include: (i) creation of unnecessary new objects with the new command, which of course takes time and reserves memory, which has to be freed later on by the garbage collection; (ii) the use of immutable objects like Strings, Integer, Long, Short, Double, Float, Character, Byte, or Boolean, which cannot be changed after creation, but are newly created upon change; and (iii) the unnecessary use of synchronized objects, i.e., Vector class, StringBuffer class, etc.

5.2   INDEXING AND SEARCHING

With LIRE, indexing and search are done with just a few lines of code. Listing 5.3 gives all the code needed to index a dataset with LIRE. First, all image files are collected with a utility class, then a DocumentBuilder instance is created. Here, we use the CEDD feature, which performs well for many use cases. After that we create an IndexWriter[6] (cp. lines 6–10). The main loop (cp. lines 11–23) iterates through all images, creates Lucene Documents and puts them into the index. Finally, the IndexWriter is closed.

The use of a custom image feature within LIRE is simplified by the GenericDocumentBuilder and the GenericImageSearcher. So if you want to adapt the code of Listing 5.3 to your custom feature, e.g., named SampleFeatureImplementation, just comment out line 3 and uncomment lines 4 and 5. Note that the String parameter, saying "fNameSampleFeature" in the example, defines the field in Lucene, where the values of your feature are stored. You have to use the same field name for search.

Indexing multiple features of one image at the same time is done with an instance of the ChainedDocumentBuilder. There you can add multiple DocumentBuilder instances of different classes with the add(DocumentBuilder) method. An example is given in Listing 5.7.

Listing 5.3: Sample code for indexing images with LIRE

```
1   ArrayList<String> images =
2           FileUtils.getAllImages(new File(args[0]), true);
3   DocumentBuilder builder = DocumentBuilderFactory.getCEDDDocumentBuilder();
4   // DocumentBuilder builder = new GenericDocumentBuilder(
5   //   SampleFeatureImplementation.class, "fNameSampleFeature");
6   IndexWriterConfig conf =
7           new IndexWriterConfig(Version.LUCENE_40,
8           new WhitespaceAnalyzer(Version.LUCENE_40));
9   IndexWriter iw = new IndexWriter(
10          FSDirectory.open(new File("index")), conf);
11  for (Iterator<String> it = images.iterator(); it.hasNext(); ) {
12      String imageFilePath = it.next();
13      System.out.println("Indexing " + imageFilePath);
14      try {
15          BufferedImage img = ImageIO.read(
16                  new FileInputStream(imageFilePath));
17          Document document =
18                  builder.createDocument(img, imageFilePath);
19          iw.addDocument(document);
20      } catch (Exception e) {
21          e.printStackTrace();
22      }
23  }
24  iw.close();
```

Searching with the query-by-example paradigm in this newly created index is also done with only a few lines of code (cp. listing 5.4). Under the assumption that we want to search our index for an image visually similar to the image stored as "0.jpg," we create a BufferedImage instance of the image file "0.jpg" and open the index at the same location we stored it. The ImageSearcher has to be of the same type as the DocumentBuilder. So, if we used a DocumentBuilder for CEDD at indexing time, we can use a ImageSearcher for CEDD at search time.[7] The actual searching is done

---

[6]Based on the latest Lucene version at the time of writing.

[7]If you use multiple features make sure that you chose the right ones at indexing time. Searching for features, that are not in the index, is a common cause of errors.

at line 6. Finally, the results are printed to System.out. The text below shows a sample output[8] and Figure 5.2 for the first five images of the result list.

```
Result of running the code in listing~\ref{listing:LireIndexingSearch}:
1.0000:   0.jpg   0.1835:    87.jpg
0.1543:  58.jpg   0.0988:   255.jpg
0.0607: 394.jpg   0.0571:     2.jpg
0.0566:  56.jpg   0.0226:   240.jpg
0.0059:  17.jpg   0.0000:    15.jpg
```

Listing 5.4: Sample code for searching images with LIRE

```
1   BufferedImage img = ImageIO.read("0.jpg");
2   IndexReader ir = DirectoryReader.open(FSDirectory.open(new File("index")));
3   ImageSearcher searcher =
4       ImageSearcherFactory.createCEDDImageSearcher(10);
5   // new GenericFastImageSearcher(SampleFeatureImplementation.class, "fNameSampleFeature");
6   ImageSearchHits hits = searcher.search(img, ir);
7   for (int i = 0; i < hits.length(); i++) {
8       String fileName = hits.doc(i).getValues(
9           DocumentBuilder.FIELD_NAME_IDENTIFIER)[0];
10      System.out.println(hits.score(i) + ": \t" + fileName);
11  }
```

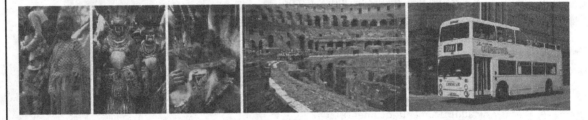

Figure 5.2: Five out of ten results from a search with the code in Listing 5.4 utilizing images from the SIMPLIcity data set [81]. The left-most image was used as an example.

Once again, integrating custom features is a matter of a single line. So if you want to adapt the code of Listing 5.4 to your custom feature, e.g., named SampleFeatureImplementation, just comment out lines 3 and 4 and uncomment line 5. Note that the DocumentBuilder interface provides all the necessary field names that are used within LIRE and Lucene, i.e., the name of the field where the path to the image is stored (cp. line 9 in Listing 5.3).

[8]The number on the left is the degree of similarity (or *score*) within the [0, 1] range between the query image (0.jpg in this case) and the image whose filename appears on the right-hand side of each line.

5.3   ADVANCED FEATURES

In addition to basic linear search with global features LIRE also provides means to handle local features, approximate indexing, and result re-ranking.

5.3.1   BAG OF VISUAL WORDS

Due to the fact that a local feature extraction algorithm usually returns multiple local descriptors per image (see also Section 3.3), the `LireFeature` interface cannot be used with local features in the same way as with global features. The first step is to extract all available local features from all the images in the dataset and put them into an index. The basic entry point for local features in LIRE is the respective `DocumentBuilder` implementation, e.g., the `SurfDocumentBuilder` takes care of the SURF point extraction using the *jopensurf*[9] library. In the second step, a vocabulary of visual words has to be created by clustering (a sample of) the local features and the visual words have to be assigned to the images (see also Section 4.4). LIRE wraps the clustering and assignment steps in their respective `LocalFeatureHistogramBuilder` class, i.e., the `SurfFeatureHistogramBuilder` in case of SURF features.

Listing 5.5 gives the code for indexing a dataset using the bag of visual words (BOVW) approach. Lines 1–13 reflect the common usage of a `DocumentBuilder` implementation. Lines 14–16 indicate the additional step of creating the visual words vocabulary and re-indexing all images by assigning visual words based on the newly created vocabulary. The `SurfFeatureHistogramBuilder` can be configured to use a fraction of all images that are in the dataset (second parameter, i.e., 1,000 in Listing 5.5) and how large the visual vocabulary should be (third parameter, i.e., 500 in Listing 5.5). At this point, note that these two parameters influence runtime and storage complexity as the number of clusters as well as the number of features to cluster are both critical to the employed k-means algorithm.

Listing 5.5: Sample code for indexing with bag of visual words in LIRE

```
1   DocumentBuilder builder = new SurfDocumentBuilder();
2   IndexWriterConfig conf = new IndexWriterConfig(Version.LUCENE_40,
3           new WhitespaceAnalyzer(Version.LUCENE_40));
4   IndexWriter iw =
5           new IndexWriter(FSDirectory.open(new File("index")), conf);
6   ArrayList<File> files =
7           FileUtils.getAllImageFiles(new File("testdata"), true);
8   for (Iterator<File> i = files.iterator(); i.hasNext(); ) {
9       File imgFile = i.next();
10      iw.addDocument(builder.createDocument(
11              ImageIO.read(imgFile), imgFile.getPath()));
12  }
13  iw.close();
14  IndexReader ir = DirectoryReader.open(FSDirectory.open(new File("index")));
15  SurfFeatureHistogramBuilder sfh =
16          new SurfFeatureHistogramBuilder(ir, 1000, 500);
17  sfh.index();
```

[9]http://code.google.com/p/jopensurf/

Searching the index based on visual words is done by using the VisualWordsImageSearcher (cp. Listing 5.6). Note that there is an additional step needed for images to create the visual words based on the vocabulary (cp. line 5).

Listing 5.6: Sample code for sea with visual words in LIRE

```
1  VisualWordsImageSearcher vis = new VisualWordsImageSearcher(100,
2      DocumentBuilder.FIELD_NAME_SURF_VISUAL_WORDS);
3  Document queryDoc =
4      builder.createDocument(queryImage, "query");
5  queryDoc = sfh.getVisualWords(queryDoc);
6  ImageSearchHits hits = vis.search(queryDoc, ir);
```

The visual words vocabulary is managed by the `SurfFeatureHistogramBuilder` class. It creates a file names "clusters-surf.dat" upon vocabulary creation and reads the vocabulary on (i) index updates with the method `indexMissing()` and (ii) the creation of visual words for single documents with `getVisualWords(Document)`. The file is stored in the working directory of the running application, so if two different vocabularies are needed, a developer either needs to extend the `SurfFeatureHistogramBuilder` or to start another instance from another working directory.

5.3.2   RESULT RE-RANKING AND FILTERING

Result re-ranking is a common practice to combine different aspects of different low-level features. If, for instance, retrieval should incorporate a specific texture feature, but users like to see color information included in the result list, one can first search based on the texture feature for the 100 best matching images and then re-rank these 100 images based on a color feature. Needless to say, the index has to include both of the features, therefore, a `ChainedDocumentBuilder` needs to be used. An example for the usage of a `ChainedDocumentBuilder` is given in Listing 5.7.

Listing 5.7: Use of the ChainedDocumentBuilder class

```
1  ChainedDocumentBuilder builder = new ChainedDocumentBuilder();
2  builder.addBuilder(DocumentBuilderFactory.getCEDDDocumentBuilder());
3  builder.addBuilder(DocumentBuilderFactory.getColorLayoutBuilder())
```

The Java source code needed for re-ranking is given in Listing 5.8. In this example, the index is searched with the CEDD feature (cp. lines 4–6). The RerankFilter implementation is created using the `ColorLayout` feature and applied on the `ImageSearchHits` instance (cp. lines 7–9). Note that you need the initial query document for filtering. The results can then be printed using the same code shown earlier in Listing 5.3 (cp. lines 6–9).

Listing 5.8: Sample code for re-ranking results with LIRE

```
1  IndexReader reader = DirectoryReader.open(
2      FSDirectory.open(new File(indexPath)));
3  Document document = reader.document(0);
4  ImageSearcher searcher =
5      ImageSearcherFactory.createCEDDImageSearcher(100);
6  ImageSearchHits hits = searcher.search(document, reader);
7  RerankFilter filter = new RerankFilter(ColorLayout.class,
```

```
8       DocumentBuilder.FIELD_NAME_COLORLAYOUT);
9   hits = filter.filter(hits, document);
```

A more advanced version of re-ranking is the LSA filter implementation. In this filter, *Latent Semantic Analysis* [18] is applied to the result set, which is known to work well in many use cases. However, LSA is quite a demanding algorithm, so the size of the initial result set is critical to the re-ranking performance. An effect of LSA filtering is shown in Figure 5.3. The original results (on the left) are re-ranked based on LSA. The modified (i.e., re-ranked) results are shown on the right. The image with the red border is the query image. In contrast to the re-ranking filter, the LSA filter does not necessarily need to operate on a feature different to the one searched for.

Listing 5.9: Sample code for using LSA with LIRE

```
1   IndexReader reader = IndexReader.open(
2       FSDirectory.open(new File(indexPath)));
3   Document document = reader.document(0);
4   ImageSearcher searcher =
5       ImageSearcherFactory.createCEDDImageSearcher(100);
6   ImageSearchHits hits = searcher.search(document, reader);
7   LsaFilter filter = new LsaFilter(CEDD.class,
8       DocumentBuilder.FIELD_NAME_CEDD);
9   hits = filter.filter(hits, document);
```

Figure 5.3: Example of LSA filtering using the ColorLayoutFeature (on the right) compared to the original results (on the left). The original best match remains the same, the second-best moves to the third spot, and the other three matches are replaced by different candidates.

Note that both filtering routines presented here can be tried in LireDemo using the *Developer* menu. Furthermore, custom implementations of filters are supported in LIRE. Developers need to implement the `SearchHitsFilter` interface with its single method `filter(ImageSearchHits, Document)`.

5.4   HOW TO APPLY LIRE

LIRE provides a handy toolbox for rapid development of a VIR system. When it is time to move from an early prototype to a full-fledged solution, however, several additional steps must be taken, among them: (i) the identification of the domain (or *scenario*), image data, and the nature of the queries; (ii) creation of a domain-specific test data set and benchmark; and (iii) field and deployment tests with actual users.

5.4.1   SCENARIO INVESTIGATION

Figure 5.4 shows two different scenarios[10] which will—most likely—correspond to different underlying datasets. While the upper row in Figure 5.4 shows examples of a potential dataset containing landscape pictures, the lower row shows examples from a dataset of macro shots.

Figure 5.4: Example photo sets. The upper row illustrates a set of landscape photos, while the other row shows examples for macro shots.

The first step for a developer would be to find out about the characteristics of the images in the dataset. The following questions can provide a guideline for investigation.

What is shown on the images? Are they artificial, natural, or heavily edited? This heavily influences the choice of the low-level features employed. Clip arts and artificial images have different

---

[10]The attentive reader could argue that the two scenarios share many aspects, and that they both could be associated with keywords such as *outdoor*, *nature*, *sky*, etc.

color distribution and contrast than natural images. The CEDD feature, for instance, works well for natural images, but might not be the descriptor of choice for artificial images.

What is the quality of the images? Are they of high quality, e.g., high contrast, or very colorful? Low-quality images, like underexposed digital photos, need to be preprocessed to provide useful input for color and texture descriptors. Moreover, compression artifacts might interfere with texture descriptors.

Is there a common theme? Do the images show similar objects, objects from the same category? If so, what can we learn about the images to make the low-level features and/or the preprocessing more meaningful? Take the two example sets in Figure 5.4 for instance. While both rows show natural images, the theme is different. If a dataset contains silhouetted images with a uniform background, the background pixels may be removed to focus at the main objects. If the dataset includes grayscale and color images of the same things, no color feature should be used. If a dataset contains buildings and architectural designs, local features like SURF and SIFT are known to work well.

What is the format of the images? What is the minimum and maximum resolution? Resolution is a critical factor for local features, i.e., the bigger the image, the more local features can be found. Also, resolution influences extraction time. In many cases, images can be scaled down to a common maximum side length of 1,000 or even less pixels, without influencing retrieval performance too much. Note that the scaling also has to be done for query images.

How many images are there? Are there thousands or millions of images? If there are only few—up to say 100,000—images, linear search is a valid option. If there are millions of images, a sub-linear indexing and search approach has to be considered. Furthermore, what is the growth rate of the dataset? Are new images being added every minute, every week, or every year? For bag of visual word indexing, for instance, index updates are always time-consuming, so they have to be carefully planned.

What will the users search for? What is the actual task of the user? Will users search for visually similar images, or will they try to find duplicates? Do they need to find cropped and altered versions of images, e.g., images from the dataset that have been used on web pages? Based on the task, the candidate low-level features must be selected. If cropped images have to be found, ColorLayout is *not* a feature of choice; one might consider local features instead. If visually similar images are needed, e.g., searching for all sunsets in the database, ColorLayout may work out well.

5.4.2   BENCHMARKING

As soon as the characteristics of the dataset and the nature of the queries are clear, a test dataset for benchmarking has to be prepared. The test dataset should be as extensive as possible and should aim to reflect the actual data well. This is typically achieved by randomly drawing a large number of samples from the original data. Typically, several rounds of indexing are involved in benchmarking, so a too large sample leads to extensive benchmarking time requirements. Depending on the heterogeneity of the actual data and the selection of features, the size of the benchmarking data set may vary from

1,000–10,000, or even more images. Along with the images, topics need to be defined. Again, we assume the more topics there are, the more significant the retrieval performance results will be.

The next step is to decide upon retrieval performance measures that are most appropriate to the topics and scenario at hand (see section 3.5 for more information). If each topic contains one result for one query image a retrieval measure such as MAP or precision at ten $p_{at}(10)$ does not make any sense. In this case, the error rate or the inverted rank are the measures of choice. If the scenario includes a mobile client on a small screen, where only three or five results can be shown at a time, $p_{at}(3)$ or $p_{at}(5)$ are good candidates. Typically, the computation of the retrieval performance does not take too much time, so multiple measures can be computed in a benchmarking suite.

Finally, a benchmarking suite should be created, where a selection of different low-level features, indexing strategies, preprocessing methods etc. can be compared to each other. A benchmarking suite should allow for quick and easy ways to change topics, datasets and employed retrieval methods. A sample result of benchmarking is shown in Table 5.1, where several low-level features implemented in LIRE have been tested on the SIMPLIcity data set [81] for publication in [49]. In this case, the best-performing low-level features CEDD, JCD, and Auto color correlogram have high mean average precision (MAP) and precision at ten $p_{at}(10)$ values, accompanied by low error rates (er).

Table 5.1: Performance of selected low-level features tested on the SIMPLIcity dataset [81] in terms of mean average precision (MAP), precision at ten ($p_{at}(10)$) and error rate (er) (taken from [49] and updated to the current LIRE version)

Feature	map	$p_{at}(10)$	er
Auto color correlogram	0.475	0.725	0.171
CEDD	0.506	0.710	0.178
Color histogram	0.450	0.704	0.191
FCTH	0.498	0.703	0.209
Gabor	0.233	0.248	0.707
JCD	0.510	0.719	0.177
Joint histogram	0.453	0.691	0.196
JPEG coefficients histogram	0.446	0.669	0.215
MPEG-7 color layout	0.439	0.610	0.309
MPEG-7 scalable color	0.305	0.470	0.462
MPEG-7 edge histogram	0.333	0.500	0.401
SIFT BoVW	0.183	0.243	0.687
Tamura	0.253	0.359	0.601

5.4.3 DEPLOYMENT TESTS AND PERFORMANCE OPTIMIZATION

The example in Table 5.1 has shown that oftentimes there is no clear winner in terms of best performing low-level features. So, ultimately, the actual users have to decide upon the actually employed methods. Therefore, a field test involving actual users should be done next. Users need to work with the proposed VIR system and should provide feedback on the quality of results.

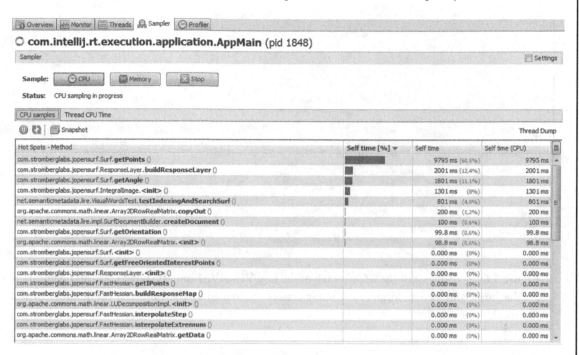

Figure 5.5: Screenshot of the Java VisualVM tool showing the *Sampler* tab, where CPU time and memory consumed at method level are listed.

Finally, when the selection of methods is done, the VIR system needs to be reviewed and fine-tuned from the perspective of runtime performance and memory consumption. The Oracle Java SDK already comes with a handy tool packaged: the Java VisualVM (to be found in the `bin` directory). As it can be seen in Figures 5.5 and 5.6, the tool shows online memory consumption, thread management, garbage collection statistics, and even provides a profiler for memory and CPU usage at class method level. With such a tool, a developer can investigate which methods take most time, generate most objects, etc., and can inspect those methods in more detail. Figure 5.5 shows statistics on the extraction of SURF points for indexing. As it can be easily seen, the method `Surf.getPoints()` takes more than 60% of the runtime and should therefore be investigated and optimized in great detail for a high-performance system. Figure 5.6 gives a snapshot of the online general statistics of the same extraction code.

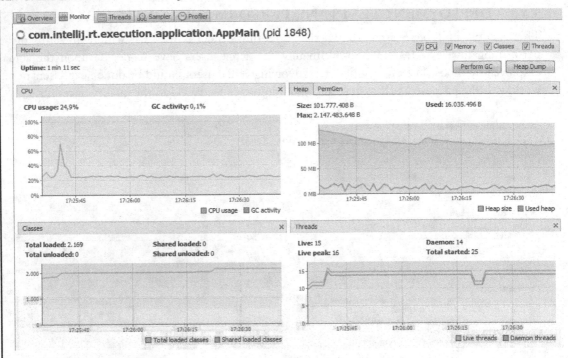

Figure 5.6: Screenshot of the *Monitor* tab of the Java VisualVM tool. It shows CPU and memory usage graphs (top), and also monitors how many classes are loaded at the time and the number of threads (bottom).

SUMMARY

In this chapter, we provided instructions, sample code, and detailed technical advice on how to use LIRE to index and search images, filter, and re-rank results, including some of LIRE's advanced methods such as bag of visual words and locality sensitive hashing. We also offered guidelines to employing LIRE in real-world scenarios.

PROBLEMS

5.1 Write a custom feature that combines rank of a pixel and color in a joint histogram with 32 color bins and 4 levels of rank.

5.2 Index and search the Ferrari test data set with LIRE and your custom-built feature using `GenericDocumentBuilder` and `GenericFastImageSearcher`.

5.3 Apply the `LSAFilter` to the results created with your custom feature to see re-ranking at work.

CHAPTER 6

Concluding Remarks

This book has taken you through a journey from basic concepts in digital imaging and information retrieval to advanced tips on how to optimize the performance of a practical visual information retrieval solution implemented in Java with the help of LIRE.

There are many exciting developments in the field of visual information retrieval that could not fit into the size and scope of this introductory, implementation-oriented text. In this chapter, we provide a glimpse into some of the most promising research directions, challenges, and opportunities in the area, as well as a list of useful resources for those who want to contribute their research efforts to the advancement of the field.

6.1 RESEARCH DIRECTIONS, CHALLENGES, AND OPPORTUNITIES

Visual information retrieval is essentially a *multidisciplinary field*, with contributions from a diverse array of well-established knowledge areas, such as: (document-based) information retrieval, image processing and computer vision, human-computer interaction, data mining, machine learning, human visual perception, and database systems, among others.

The principles and techniques discussed in this book can be applied to a variety of *specialized domains*, such as: medical information systems, security (e.g., biometrics), personal photo collections, public image repositories (e.g., Flickr, Facebook), museum archives, and many others. Moreover, VIR solutions can be deployed in a number of *different platforms*, from desktop applications to web-based systems to mobile apps.

More than 10 years ago, Smeulders and colleagues stated content-based image retrieval "will continue to grow in every direction: new audiences, new purposes, new styles of use, new modes of interaction, larger data sets, and new methods to solve the problems." [74] We believe this is still true. Some of the most exciting new avenues for visual information retrieval lie at the intersection of some of these fields, domains, and platforms. We have selected three of them, based on our recent and ongoing research efforts.

Mobile Visual Search (MVS)
Mobile visual search provides the capability to perform content-based image retrieval from a mobile device. It has experienced an enormous growth in recent years thanks to: (i) the increasing popularity and affordability of smartphones and tablets with powerful CPUs, high-quality cameras, GPS, and high-speed Internet access capabilities; and (ii) the dramatic increase in mobile photography camera

culture and associated popularization of mobile apps for image processing, filtering, and sharing, such as Facebook or Instagram. Perhaps more importantly (from the point of view of this book) is that mobile visual search finally provides a natural case for the query-by-example (QBE) paradigm, after all—in most cases—the example is right in front of the user! Snap a picture, press a button, and voilà! Similar results (or augmented information, e.g., product price and description) appear on the mobile device screen.

Some of the technical challenges include [25]: (i) deciding how much of the processing should be performed on the client and how much should be server-based; (ii) optimizing performance so that the client-side processing is fast and economical in terms of battery consumption; (iii) keeping the amount of data transmitted over the network to a minimum (which is associated with the ongoing research topic of finding compact image descriptors for MVS applications); and (iv) designing and implementing robust feature extraction and image representation algorithms, capable of providing reliable recognition of objects captured under a wide range of conditions, e.g., varying distances, viewing angles, and lighting conditions, partial occlusions or motion blur.

The opportunities are endless and the field is not yet overcrowded. Despite the early success of Google Goggles and other comparable products in demonstrating the basic proof of concept, most contemporary solutions only work on narrow domains (e.g., book covers, wine labels, and company logos) and/or require extensive training and/or depend on always-on Internet connectivity.

Medical image retrieval

The field of *Content-Based Medical Image Retrieval* (CBMIR) holds great promise of growth and commercial success. After all, medical professionals, hospitals, and clinics use images extensively, deal with domains that are (almost by definition) narrow, and amass significant expert knowledge that might be properly tapped to evaluate, refine, and improve VIR solutions.

Some of the challenges behind medical image retrieval include: the need to comply with different standards and terminology (e.g., DICOM [63] and the IRMA code [45, 64]), and modality dependencies (a technique that works well for chest x-rays is not guaranteed to succeed for brain MRI). Moreover, there are equipment dependencies (which, if encoded as metadata, can actually improve the functionality and quality of results of a medical VIR system), and issues of privacy and proprietary data, which are not present in large-scale web-based image retrieval for instance.

The opportunities in medical image retrieval are many, ranging from the creation of better user interfaces (responsive, highly interactive, and capable of supporting relevance feedback) to new applications of CBMIR in medical teaching, research, and diagnosis, to extension of existing systems to new devices, such as tablets and smartphones.

Human computation, crowdsourcing, and serious games

The multimedia research community has recently started to employ human computation methods [44] in an attempt to bridge—or at least narrow—the *semantic gap* (the difference between the data that can be captured from raw pixels and the high-level interpretation assigned by humans to the associated images).

Crowdsourcing tools, such as Amazon Mechanical Turk, have become quite popular methods for obtaining human data in tasks associated to the theme of this book, e.g., image tagging, subjective analysis of image quality, and image matching. An alternative approach to tap onto human intelligence consists in designing and deploying serious games, or "games with a purpose (GWAPs)" [80]. The implementation of a well-designed GWAP that provides an enjoyable gaming experience with a social component has the potential to unlock a large source of crowdsourcing knowledge for advancing the state of the art in image retrieval research.

6.2   RESOURCES

These are some of the magazines and journals that publish research results in visual information retrieval and related areas (in no particular order).

- *ACM Multimedia Systems*

- *ACM Transactions on Multimedia Computing, Communications, and Applications (TOMCCAP)*

- *Computer Vision and Image Understanding (CVIU)*

- *EURASIP Journal on Image and Video Processing*

- *IEEE Multimedia Magazine*

- *IEEE Transactions on Circuits and Systems for Video Technology (CSVT)*

- *IEEE Transactions on Image Processing (TIP)*

- *IEEE Transactions on Medical Imaging*

- *IEEE Transactions on Multimedia*

- *IEEE Transactions on Pattern Analysis and Machine Intelligence (PAMI), Image, and Vision Computing*

- *International Journal of Computer Vision (IJCV)*

- *Journal of Visual Communication and Image Representation*

- *Multimedia Tools and Applications (MTAP)*

- *Pattern Recognition*

- *Signal Processing: Image Communication*

Additionally, there are a number of conferences in which advances in this area are systematically reported, among them (in no particular order):

- *ACM International Conference on Multimedia (ACM MM)*

- *ACM International Conference on Multimedia Retrieval (ICMR)*

- *ACM Special Interest Group on Information Retrieval Conference (SIGIR)*

- *IEEE Computer Vision and Pattern Recognition Conference (CVPR)*

- *IEEE International Conference on Multimedia and Expo (ICME)*

- *IEEE International Symposium on Multimedia (ISM)*

- *International Workshop on Image Analysis for Multimedia Interactive Services (WIAMIS).*

Bibliography

[1] Apache lucene project. http://lucene.apache.org/. Cited on page(s) 8

[2] Luke (lucene index toolbox). http://code.google.com/p/luke/. Cited on page(s) 7

[3] Secure hash standard (shs), March 2012. Cited on page(s) 58

[4] J. Allan, B. Croft, A. Moffat, and M. S. (Eds.). Frontiers, challenges and opportunities for information retrieval: Report from swirl 2012. *ACM SIGIR Forum*, 46:2–32, 2012. DOI: 10.1145/2215676.2215678 Cited on page(s) 22

[5] G. Amato and P. Savino. Approximate similarity search in metric spaces using inverted files. In *Proceedings of the 3rd international conference on Scalable information systems*, InfoScale '08, pages 28:1–28:10, ICST, Brussels, Belgium, Belgium, 2008. ICST (Institute for Computer Sciences, Social-Informatics and Telecommunications Engineering). DOI: 10.4108/ICST.INFOSCALE2008.3486 Cited on page(s) 60, 61

[6] R. Baeza-Yates and B. Ribeiro-Neto. *Modern Information Retrieval*. Addison-Wesley/ACM Press, New York, 1999. Cited on page(s) 14, 17, 19, 22

[7] H. Bay, T. Tuytelaars, and L. Van Gool. Surf: Speeded up robust features. *Computer Vision— ECCV 2006*, pages 404–417, 2006. DOI: 10.1007/11744023_32 Cited on page(s) 42

[8] N. Beckmann, H.-P. Kriegel, R. Schneider, and B. Seeger. The r*-tree: an efficient and robust access method for points and rectangles. In *Proceedings of the 1990 ACM SIGMOD international conference on Management of data*, SIGMOD '90, pages 322–331, New York, NY, USA, 1990. ACM. DOI: 10.1145/93605.98741 Cited on page(s) 55

[9] C. Böhm, S. Berchtold, and D. Keim. Searching in high-dimensional spaces: Index structures for improving the performance of multimedia databases. *ACM Computing Surveys (CSUR)*, 33(3):322–373, 2001. DOI: 10.1145/502807.502809 Cited on page(s) 55

[10] G. Box and M. Muller. A note on the generation of random normal deviates. *The Annals of Mathematical Statistics*, 29(2):610–611, 1958. DOI: 10.1214/aoms/1177706645 Cited on page(s) 58

[11] S. Brin and L. Page. The anatomy of a large-scale hypertextual web search engine. *Computer networks and ISDN systems*, 30(1):107–117, 1998. DOI: 10.1016/S0169-7552(98)00110-X Cited on page(s) 22

[12] W. Burger and M. J. Burge. *Digital Image Processing: an algorithmic introduction using Java.* Springer, New York, 2008. Cited on page(s) 33

[13] T. Chang and C.-C. Jay Kuo. Texture analysis and classification with tree-structured wavelet transform. *IEEE Transactions on Image Processing*, 2(4), 1993. DOI: 10.1109/83.242353 Cited on page(s) 37

[14] S. Chatzichristofis and Y. Boutalis. Cedd: Color and edge directivity descriptor: A compact descriptor for image indexing and retrieval. In A. Gasteratos, M. Vincze, and J. Tsotsos, editors, *Computer Vision Systems*, volume 5008 of *Lecture Notes in Computer Science*, pages 312–322. Springer Berlin / Heidelberg, 2008. Cited on page(s) 35, 36, 40, 44

[15] S. Chatzichristofis and Y. Boutalis. Fcth: Fuzzy color and texture histogram - a low level feature for accurate image retrieval. In *Image Analysis for Multimedia Interactive Services, 2008. WIAMIS '08. Ninth International Workshop on*, pages 191 –196, may 2008. DOI: 10.1109/WIAMIS.2008.24 Cited on page(s) 40, 44

[16] S. Chatzichristofis, Y. Boutalis, and M. Lux. Selection of the proper compact composite descriptor for improving content based image retrieval. In *Proceedings of the 6th IASTED International Conference*, volume 134643, page 064, 2009. Cited on page(s) 40

[17] M. Datar, N. Immorlica, P. Indyk, and V. S. Mirrokni. Locality-sensitive hashing scheme based on p-stable distributions. In *Proceedings of the twentieth annual symposium on Computational geometry*, SCG '04, pages 253–262, New York, NY, USA, 2004. ACM. DOI: 10.1145/997817.997857 Cited on page(s) 58

[18] S. Deerwester, S. T. Dumais, G. W. Furnas, T. K. Landauer, and R. Harshman. Indexing by latent semantic analysis. *JOURNAL OF THE AMERICAN SOCIETY FOR INFORMATION SCIENCE*, 41(6):391–407, 1990. DOI: 10.1002/(SICI)1097-4571(199009)41:6%3C391::AID-ASI1%3E3.0.CO;2-9 Cited on page(s) 22, 77

[19] T. Deselaers, D. Keysers, and H. Ney. Features for image retrieval: an experimental comparison. *Information Retrieval*, 11:77–107, 2008. DOI: 10.1007/s10791-007-9039-3 Cited on page(s) 34, 38, 44, 46

[20] T. Deserno, S. Antani, and R. Long. Ontology of gaps in content-based image retrieval. *Journal of Digital Imaging*, 22:202–215, 2009. DOI: 10.1007/s10278-007-9092-x Cited on page(s) 6

[21] P. Diaconis. *Group representations in probability and statistics.* Lecture notes-monograph series. Institute of Mathematical Statistics, 1988. Cited on page(s) 60

[22] H. Eidenberger. Distance measures for mpeg-7-based retrieval. In *Proceedings of the 5th ACM SIGMM international workshop on Multimedia information retrieval*, MIR '03, pages 130–137, New York, NY, USA, 2003. ACM. DOI: 10.1145/973264.973286 Cited on page(s) 44

[23] P. Enser and C. Sandom. Towards a comprehensive survey of the semantic gap in visual image retrieval. In *Proceedings of the 2nd international conference on Image and video retrieval*, CIVR'03, pages 291–299, Berlin, Heidelberg, 2003. Springer-Verlag. DOI: 10.1007/3-540-45113-7_29 Cited on page(s) 6

[24] C. Faloutsos and K.-I. Lin. Fastmap: a fast algorithm for indexing, data-mining and visualization of traditional and multimedia datasets. *SIGMOD Rec.*, 24(2):163–174, May 1995. DOI: 10.1145/223784.223812 Cited on page(s) 55, 56

[25] B. Girod, V. Chandrasekhar, D. Chen, N.-M. Cheung, R. Grzeszczuk, Y. Reznik, G. Takacs, S. Tsai, and R. Vedantham. Mobile visual search. *Signal Processing Magazine, IEEE*, 28(4):61–76, july 2011. DOI: 10.1109/MSP.2011.940881 Cited on page(s) 84

[26] E. Goldstein. *Sensation & Perception*. Brooks/Cole, Pacific Grove, CA, 1999. Cited on page(s) 31

[27] R. C. Gonzalez and R. E. Woods. *Digital Image Processing*. Prentice-Hall, Upper Saddle River, NJ, 2nd edition, 2002. Cited on page(s) 37

[28] M. H. Gross, R. Koch, L. Lippert, and A. Dreger. Multiscale image texture analysis in wavelet spaces. In *Proc. IEEE International Conference on Image Processing*, volume 3, pages 412–416, Austin, TX, USA, November 1994. DOI: 10.1109/ICIP.1994.413816 Cited on page(s) 37

[29] A. Guttman. R-trees: a dynamic index structure for spatial searching. *SIGMOD Rec.*, 14(2):47–57, June 1984. DOI: 10.1145/602259.602266 Cited on page(s) 55

[30] M. Hall, E. Frank, G. Holmes, B. Pfahringer, P. Reutemann, and I. Witten. The weka data mining software: an update. *ACM SIGKDD Explorations Newsletter*, 11(1):10–18, 2009. DOI: 10.1145/1656274.1656278 Cited on page(s) 65

[31] A. Hanjalic, C. Kofler, and M. Larson. Intent and its discontents: the user at the wheel of the online video search engine. In *Proceedings of the 20th ACM international conference on Multimedia*, MM '12, pages 1239–1248, New York, NY, USA, 2012. ACM. DOI: 10.1145/2393347.2396424 Cited on page(s) 6

[32] R. Haralick, K. Shanmugam, and I. Dinstein. Texture features for image classification. *IEEE Transactions on Systems, Man, and Cybernetics*, 3(6), 1973. DOI: 10.1109/TSMC.1973.4309314 Cited on page(s) 37

[33] J. S. Hare, P. H. Lewis, P. G. B. Enser, and C. J. Sandom. Mind the gap: another look at the problem of the semantic gap in image retrieval. pages 607309–607309–12, 2006. DOI: 10.1117/12.647755 Cited on page(s) 6

[34] J. S. Hare, S. Samangooei, and D. P. Dupplaw. Openimaj and imageterrier: Java libraries and tools for scalable multimedia analysis and indexing of images. In *Proceedings of the 19th ACM international conference on Multimedia*, MM '11, pages 691–694, New York, NY, USA, 2011. ACM. DOI: 10.1145/2072298.2072421 Cited on page(s) 64

[35] J. Huang, S. Kumar, M. Mitra, W.-J. Zhu, and R. Zabih. Image indexing using color correlograms. In *Proc. IEEE International Conference on Computer Vision and Pattern Recognition*, pages 762–768, San Juan, Puerto Rico, 1997. DOI: 10.1109/CVPR.1997.609412 Cited on page(s) 39, 40

[36] M. J. Huiskes and M. S. Lew. The mir flickr retrieval evaluation. In *MIR '08: Proceedings of the 2008 ACM International Conference on Multimedia Information Retrieval*, New York, NY, USA, 2008. ACM. DOI: 10.1145/1460096.1460104 Cited on page(s) 46

[37] M. J. Huiskes, B. Thomee, and M. S. Lew. New trends and ideas in visual concept detection: The mir flickr retrieval evaluation initiative. In *MIR '10: Proceedings of the 2010 ACM International Conference on Multimedia Information Retrieval*, pages 527–536, New York, NY, USA, 2010. ACM. DOI: 10.1145/1743384.1743475 Cited on page(s) 46

[38] A. K. Jain, M. N. Murty, and P. J. Flynn. Data clustering: a review. *ACM Comput. Surv.*, 31(3):264–323, Sept. 1999. DOI: 10.1145/331499.331504 Cited on page(s) 64

[39] C. Kofler and M. Lux. Dynamic presentation adaptation based on user intent classification. In *Proceedings of the 17th ACM international conference on Multimedia*, MM '09, pages 1117–1118, New York, NY, USA, 2009. ACM. DOI: 10.1145/1631272.1631526 Cited on page(s) 6

[40] M. Kogler and M. Lux. Bag of visual words revisited: an exploratory study on robust image retrieval exploiting fuzzy codebooks. In *Proceedings of the Tenth International Workshop on Multimedia Data Mining*, MDMKDD '10, pages 3:1–3:6, New York, NY, USA, 2010. ACM. DOI: 10.1145/1814245.1814248 Cited on page(s) 64

[41] A. Laine and J. Fan. Texture classification by wavelet packet signatures. *IEEE Transactions on Pattern Analysis and Machine Intelligence*, 15(11), 1993. DOI: 10.1109/34.244679 Cited on page(s) 37

[42] A. Langville and C. Meyer. *Google's PageRank and beyond: The science of search engine rankings*. Princeton University Press, 2009. Cited on page(s) 13, 22

[43] M. Larson, M. Soleymani, M. Eskevich, P. Serdyukov, R. Ordelman, and G. Jones. The community and the crowd: Multimedia benchmark dataset development. *MultiMedia, IEEE*, 19(3):15 –23, july-sept. 2012. DOI: 10.1109/MMUL.2012.27 Cited on page(s) 46

[44] E. Law and L. von Ahn. *Human Computation*. Synthesis Lectures on Artificial Intelligence and Machine Learning. Morgan & Claypool Publishers, 2011. DOI: 10.2200/S00371ED1V01Y201107AIM013 Cited on page(s) 84

[45] T. Lehmann, H. Schubert, D. Keysers, M. Kohnen, and B. Wein. The irma code for unique classification of medical images. In *Medical Imaging 2003*, pages 440–451. International Society for Optics and Photonics, 2003. DOI: 10.1117/12.480677 Cited on page(s) 84

[46] J. Li and J. Wang. Automatic linguistic indexing of pictures by a statistical modeling approach. *Pattern Analysis and Machine Intelligence, IEEE Transactions on*, 25(9):1075–1088, 2003. DOI: 10.1109/TPAMI.2003.1227984 Cited on page(s) 46

[47] C. Liu, S. Jiang, and Q. Huang. Personalized online video recommendation by neighborhood score propagation based global ranking. In *Proceedings of the First International Conference on Internet Multimedia Computing and Service*, ICIMCS '09, pages 244–253, New York, NY, USA, 2009. ACM. DOI: 10.1145/1734605.1734661 Cited on page(s) 2

[48] D. G. Lowe. Distinctive image features from scale-invariant keypoints. *Int. J. Comput. Vision*, 60(2):91–110, Nov. 2004. DOI: 10.1023/B:VISI.0000029664.99615.94 Cited on page(s) 41

[49] M. Lux. Content based image retrieval with lire. In *Proceedings of the 19th ACM international conference on Multimedia*, pages 735–738. ACM, 2011. DOI: 10.1145/2072298.2072432 Cited on page(s) 47, 80

[50] M. Lux and S. A. Chatzichristofis. Lire: lucene image retrieval: an extensible java cbir library. In *Proceedings of the 16th ACM international conference on Multimedia*, MM '08, pages 1085–1088, New York, NY, USA, 2008. ACM. DOI: 10.1145/1459359.1459577 Cited on page(s) 8, 47

[51] M. Lux, C. Kofler, and O. Marques. A classification scheme for user intentions in image search. In *CHI '10 Extended Abstracts on Human Factors in Computing Systems*, CHI EA '10, pages 3913–3918, New York, NY, USA, 2010. ACM. DOI: 10.1145/1753846.1754078 Cited on page(s) 6

[52] M. Lux, M. Taschwer, and O. Marques. A closer look at photographers' intentions: a test dataset. In *Proceedings of the International ACM Workshop on Crowdsourcing for Multimedia held in conjunction with ACM Multimedia 2012*, Nara, Japan, 2012. DOI: 10.1145/2390803.2390811 Cited on page(s) 6

[53] B. Manjunath, J. Ohm, V. Vasudevan, and A. Yamada. Color and texture descriptors. *Circuits and Systems for Video Technology, IEEE Transactions on*, 11(6):703–715, 2001. DOI: 10.1109/76.927424 Cited on page(s) 45

[54] O. Marques. *Practical Image and Video Processing Using MATLAB*. Wiley-IEEE Press, 2011. DOI: 10.1002/9781118093467 Cited on page(s) 28, 31

[55] M. McCandless, E. Hatcher, and O. Gospodnetic. *Lucene in Action, Second Edition*. Manning, 2010. Cited on page(s) 21, 22

[56] T. Mei, B. Yang, X.-S. Hua, L. Yang, S.-Q. Yang, and S. Li. Videoreach: an online video recommendation system. In *Proceedings of the 30th annual international ACM SIGIR conference on Research and development in information retrieval*, SIGIR '07, pages 767–768, New York, NY, USA, 2007. ACM. DOI: 10.1145/1277741.1277899 Cited on page(s) 2

[57] K. Mikolajczyk and C. Schmid. A performance evaluation of local descriptors. *Pattern Analysis and Machine Intelligence, IEEE Transactions on*, 27(10):1615–1630, 2005. DOI: 10.1109/TPAMI.2005.188 Cited on page(s) 41

[58] K. Mikolajczyk, T. Tuytelaars, C. Schmid, A. Zisserman, J. Matas, F. Schaffalitzky, T. Kadir, and L. Gool. A comparison of affine region detectors. *International journal of computer vision*, 65(1):43–72, 2005. DOI: 10.1007/s11263-005-3848-x Cited on page(s) 41

[59] C. Mooers. *The theory of digital handling of non-numerical information and its implications to machine economics.* Zator Company, 1950. Cited on page(s) 13

[60] H. Müller, P. Clough, T. Deselaers, and B. Caputo. *ImageCLEF: Experimental Evaluation in Visual Information Retrieval.* Springer Publishing Company, Incorporated, 1st edition, 2010. Cited on page(s) 47

[61] J. Park, S.-J. Lee, S.-J. Lee, K. Kim, B.-S. Chung, and Y.-K. Lee. Online video recommendation through tag-cloud aggregation. *IEEE MultiMedia*, 18(1):78–87, Jan. 2011. DOI: 10.1109/MMUL.2010.6 Cited on page(s) 2

[62] G. Pass and R. Zabih. Comparing images using joint histograms. *Multimedia Systems*, 7:234–240, 1999. DOI: 10.1007/s005300050125 Cited on page(s) 39

[63] O. Pianykh. *Digital Imaging and Communications in Medicine (DICOM): A practical introduction and survival guide.* Springer, 2011. DOI: 10.2967/jnumed.109.064592 Cited on page(s) 84

[64] T. PIESCH, H. MÜLLER, C. KUHL, and T. DESERNO. Irma code ii: Unique annotation of medical images for access and retrieval. *Studies in health technology and informatics*, 180:159, 2012. DOI: 10.3233/978-1-61499-101-4-159 Cited on page(s) 84

[65] S. Prince. *Computer Vision: Models, Learning, and Inference.* Computer Vision: Models, Learning, and Inference. Cambridge University Press, 2012. DOI: 10.1017/CBO9780511996504 Cited on page(s) 41

[66] R. Rivest. The md5 message-digest algorithm, April 1992. Cited on page(s) 58

[67] S. Robertson and S. Walker. Some simple effective approximations to the 2-poisson model for probabilistic weighted retrieval. In *Proceedings of the 17th annual international ACM SIGIR conference on Research and development in information retrieval*, pages 232–241. Springer-Verlag New York, Inc., 1994. Cited on page(s) 22

[68] S. Robertson, H. Zaragoza, and M. Taylor. Simple bm25 extension to multiple weighted fields. In *Proceedings of the thirteenth ACM international conference on Information and knowledge management*, pages 42–49. ACM, 2004. DOI: 10.1145/1031171.1031181 Cited on page(s) 22

[69] Y. Rubner, C. Tomasi, and L. Guibas. The earth mover's distance as a metric for image retrieval. *International Journal of Computer Vision*, 40(2):99–121, 2000. DOI: 10.1023/A:1026543900054 Cited on page(s) 44

[70] S. Rüger. *Multimedia Information Retrieval*, volume 10 of *Synthesis lectures on information concepts, retrieval, and services*. Morgan & Claypool Publishers, 2009. DOI: 10.2200/S00244ED1V01Y200912ICR010 Cited on page(s) 58

[71] L. Shapiro and G. Stockman. *Computer Vision*. Prentice-Hall, Upper Saddle River, NJ, 2001. Cited on page(s) 37

[72] T. Sikora. The mpeg-7 visual standard for content description-an overview. *Circuits and Systems for Video Technology, IEEE Transactions on*, 11(6):696–702, jun 2001. DOI: 10.1109/76.927422 Cited on page(s) 38

[73] J. Sivic and A. Zisserman. Video google: A text retrieval approach to object matching in videos. *Computer Vision, IEEE International Conference on*, 2:1470, 2003. DOI: 10.1109/ICCV.2003.1238663 Cited on page(s) 62

[74] A. W. M. Smeulders, M. Worring, S. Santini, A. Gupta, and R. Jain. Content-based image retrieval at the end of the early years. *IEEE Trans. Pattern Anal. Mach. Intell.*, 22(12):1349–1380, Dec. 2000. DOI: 10.1109/34.895972 Cited on page(s) 5, 6, 83

[75] H. Tamura, S. Mori, and T. Yamawaki. Texture features corresponding to visual perception. *IEEE Transactions on Systems, Man and Cybernetics*, 8(6):460–473, 1978. DOI: 10.1109/TSMC.1978.4309999 Cited on page(s) 37, 38

[76] T. Tuytelaars and K. Mikolajczyk. Local invariant feature detectors: a survey. *Found. Trends. Comput. Graph. Vis.*, 3(3):177–280, July 2008. DOI: 10.1561/0600000017 Cited on page(s) 40, 41

[77] S. E. Umbaugh. *Computer Imaging: Digital Image Analysis and Processing*. CRC Press, Boca Raton, FL, 2005. Cited on page(s) 25

[78] K. E. A. van de Sande, T. Gevers, and C. G. M. Snoek. Evaluating color descriptors for object and scene recognition. *IEEE Transactions on Pattern Analysis and Machine Intelligence*, 32(9):1582–1596, 2010. DOI: 10.1109/TPAMI.2009.154 Cited on page(s) 32, 33

[79] C. van Rijsbergen. *Information Retrieval*. Butterworths, 1979. Cited on page(s) 14

[80] L. von Ahn and L. Dabbish. Designing games with a purpose. *Commun. ACM*, 51(8):58–67, 2008. DOI: 10.1145/1378704.1378719 Cited on page(s) 85

[81] J. Wang, J. Li, and G. Wiederhold. Simplicity: Semantics-sensitive integrated matching for picture libraries. *Pattern Analysis and Machine Intelligence, IEEE Transactions on*, 23(9):947–963, 2001. DOI: 10.1109/34.955109 Cited on page(s) 45, 46, 74, 80

[82] B. Yang, T. Mei, X.-S. Hua, L. Yang, S.-Q. Yang, and M. Li. Online video recommendation based on multimodal fusion and relevance feedback. In *Proceedings of the 6th ACM international conference on Image and video retrieval*, CIVR '07, pages 73–80, New York, NY, USA, 2007. ACM. DOI: 10.1145/1282280.1282290 Cited on page(s) 2

[83] X. Zhao, G. Li, M. Wang, J. Yuan, Z.-J. Zha, Z. Li, and T.-S. Chua. Integrating rich information for video recommendation with multi-task rank aggregation. In *Proceedings of the 19th ACM international conference on Multimedia*, MM '11, pages 1521–1524, New York, NY, USA, 2011. ACM. DOI: 10.1145/2072298.2072055 Cited on page(s) 2

Authors' Biographies

MATHIAS LUX

Mathias Lux is a Senior Assistant Professor at the Institute for Information Technology (ITEC) at Klagenfurt University, where he has been since 2006. He received his M.S. in Mathematics in 2004 and his Ph.D. in Telematics in 2006 from Graz University of Technology. Before joining Klagenfurt University, he worked in industry on web-based applications, as a junior researcher at a research center for knowledge-based applications, and as research and teaching assistant at the Knowledge Management Institute (KMI) of Graz University of Technology.

In research, he is working on user intentions in multimedia retrieval and production, visual information retrieval, and serious games. In his scientific career he has (co-) authored more than 60 scientific publications, has served in multiple program committees and as reviewer of international conferences, journals, and magazines, and has organized several scientific events. He is also well known for managing the development of the award-winning and popular open source tools *Caliph & Emir* and LIRE for visual information retrieval.

OGE MARQUES

Oge Marques is an Associate Professor in the Department of Computer & Electrical Engineering and Computer Science at Florida Atlantic University (FAU) (Boca Raton, Florida). He received his Ph.D. in Computer Engineering from FAU in 2001.

He has more than 20 years of teaching and research experience in the fields of image processing and computer vision, in different countries (U.S., Austria, Brazil, Netherlands, Spain, France, and India), languages (English, Portuguese, Spanish), and capacities. He is the (co-) author of more than 50 refereed journal and conference papers and several books in these topics, including the textbook *Practical Image and Video Processing Using MATLAB* (Wiley, 2011).

His research interests are in the area of *intelligent processing of visual information*, which combines the fields of image processing, computer vision, image retrieval, machine learning, serious games, and human visual perception. He is particularly interested in the combination of human computation and machine learning techniques to solve computer vision problems.

He is a senior member of both the ACM and IEEE, and a member of the IEEE Computer Society, IEEE Education Society, IEEE Signal Processing Society, and the honor societies of Tau Beta Pi, Sigma Xi, Phi Kappa Phi, and Upsilon Pi Epsilon.